Plant Cell Monographs
Volume 14

Series Editor: David G. Robinson
Heidelberg, Germany

Plant Cell Monographs

Recently Published Titles

Iris Meier
Editor

Functional Organization of the Plant Nucleus

 Springer

Editor
Prof. Dr. Iris Meier
Dept. of Plant Cellular and Molecular Biology
The Ohio State University
520 Aronoff Laboratory
Columbus, OH 43210
USA

Series Editor
Prof. Dr. David G. Robinson
Ruprecht-Karls-University of Heidelberg
Heidelberger Institute for Plant Sciences (HIP)
Department Cell Biology
Im Neuenheimer Feld 230
D-69120 Heidelberg
Germany

ISBN: 978-3-540-71057-8   e-ISBN: 978-3-540-71058-5
DOI: 10.1007/978-3-540-71058-5

Library of Congress Control Number: 2008939145

Cover design: WMX Design GmbH, Heidelberg, Germany

Printed on acid-free paper

9 8 7 6 5 4 3 2 1 0

springer.com

Editor

Iris Meier is a professor in plant cellular and molecular biology at The Ohio State University, USA. She obtained her B.S./M.S. degree in biology from the Technical University of Darmstadt, Germany, and her Ph.D. in molecular biology from the University of Düsseldorf, Germany. After a postdoctoral training at the Max Planck Institute for Plant Breeding in Cologne, Germany, and the University of California in Berkeley, USA, she has been a principal investigator at The University of Hamburg, Germany, the Dupont Experimental Station, Wilmington, DE, and since 1999 at The Ohio State University. Since 2006, she has been an associate editor for Plant Molecular Biology. Her primary research interests are in the three areas of Ran signal transduction, nuclear pore and nuclear envelope protein function, and the structure and function of long coiled coil proteins in plants.

Preface

In a presentation to the Linnean Society of London in November 1831, the Scottish botanist Robert Brown (perhaps better known for his discovery of Brownian motion) mentioned almost as an afterthought that in orchid epidermal cells, a single "circular areola" could be seen, a "nucleus of the cell as perhaps it might be termed." Thus, the term "nucleus" (from Latin nucleus or nuculeus, "little nut" or kernel) was born for the compartment of the eukaryotic cell that contains the majority of genetic information.

One hundred and seventy-seven years later, we know that the nucleus is the site where genetic information is stored in the form of DNA, and where it is protected from damage, duplicated, divided, recombined, repaired, and "expressed." For the latter, the genetic information is faithfully transcribed from DNA to RNA, then released from the nucleus into the surrounding cytoplasm. Most likely translated into polypeptide chains, the information re-enters the nucleus in the form of diverse proteins that function in the processes listed above.

These fundamental events of life rely on a multitude of tightly interconnected processes which make the nucleus by far the most complex organelle. The nucleus is surrounded by the nuclear envelope, a double membrane enclosing the entire organelle and separating its contents from the cellular cytoplasm. This membrane is impermeable to most molecules; nuclear pores, therefore, are necessary to allow for the controlled movement of molecules across the envelope. Nuclear transport is fundamental to all cell function, as movement through the pores is required for both nuclear and cytoplasmic transactions.

While the nucleus was first identified in a plant, we know now far less about the plant nucleus than we know about its equivalent in animals or in fungi, because the field of nuclear biology has been predominantly driven by the non plant model systems. More recently, however, plant biology has begun to catch up as research groups worldwide actively address the processes that define the plant nucleus. As in other areas of molecular cell biology, it is becoming increasingly evident that spatial organization of nuclear processes is also of crucial importance; in other words, for many processes the question "where?" is as important as the question "how?"

I would like to thank all the contributors for having risen to this editor's challenge to "envision as the overarching theme of the book the integration of mechanistic and organizational aspects of nuclear biology" and for having delivered timely, rich, thought-provoking, and highly informative contributions on their respective topics. I would also like to acknowledge the series editor Dr. David Robinson for his invitation to edit this volume, as well as Dr. Christina Eckey and Anette Lindquist at the Springer editorial office for their great assistance and tireless help in hunting down yet another email exchange I had misplaced. I also wish to thank the National Science Foundation for generously supporting my own research in this field.

September 2008   Iris Meier

Contents

Contributors

George C. Allen
Department of Horticultural Science, North Carolina State University, 1203 Partners 2, Campus Box 7550, Raleigh, NC 27606-7550, USA
george_allen@ncsu.edu

Mery Dafny-Yelin
Department of Molecular, Cellular and Developmental Biology, The University of Michigan, Ann Arbor, MI 48109, USA

David E. Evans
School of Life Sciences, Oxford Brookes University, Headington Campus, Oxford, OX3 0BP, UK, deevans@brookes.ac.uk

Paul Fransz
Swammerdam Institute for Life Sciences, University of Amsterdam, Amsterdam, The Netherlands
p.f.fransz@uva.nl

Katja Graumann
School of Life Sciences, Oxford Brookes University, Headington Campus, Oxford, OX3 0BP, UK

Sarah L. Irons
School of Life Sciences, Oxford Brookes University, Headington Campus, Oxford, OX3 0BP, UK

Iris Meier
Department of Plant Cellular and Molecular Biology and Plant Biotechnology Center, Ohio State University, 244 Rightmire Hall, 1060 Carmack Road, Columbus, OH 43214, USA
meier.56@osu.edu

Thomas Merkle
Institute of Genome Research and Systems Biology, Faculty of Biology III, University of Bielefeld, 33594 Bielefeld, Germany
tmerkle@cebitec.uni-bielefeld.de

Susana Moreno Díaz de la Espina
Laboratorio de Matriz Nuclear, Departamento Biología de Plantas, Centro Investigaciones Biológicas, CSIC, Ramiro de Maeztu 9, 28040 Madrid, Spain
smoreno@cib.csic.es

Annkatrin Rose
Department of Biology, Appalachian State University, 572 Rivers Street, Boone, NC 28608, USA
rosea@appstate.edu

John Runions
School of Life Sciences, Oxford Brookes University, Headington Campus, Oxford, OX3 0BP, UK

Andriy Tovkach
Department of Molecular, Cellular and Developmental Biology, The University of Michigan, Ann Arbor, MI 48109, USA

Tzvi Tzfira
Department of Molecular, Cellular and Developmental Biology, The University of Michigan, Ann Arbor, MI 48109, USA
ttzfira@umich.edu

Functional Organization of the Plant Nucleus

Iris Meier

Abstract For many years, the molecular processes in the eukaryotic nucleus were believed to occur randomly within the nuclear space, without functionally important spatial organization. Recently, much progress has been made in establishing that nuclear spatial organization is indeed important, and that many nuclear processes deserve the question "where?" as much as the question "how?" While the eukaryotic nucleus was first discovered in plants, the predominant progress in the field of its functional organization has been made in non-plant models. Recently, however, plant molecular biologists have begun to address the connection between structure and function and the integration of mechanistic and organizational aspects of the plant nucleus, too. While some aspects, such as the histone code, appear to be highly conserved between plants and animals, others, such as the composition of the nuclear pore, appear to differ profoundly. Here, we address various aspects of plant nuclear biology with the specific emphasis on the relationship between structure and function. We compare and contrast findings from plants and non-plant model systems and point out exciting future research directions.

1 Introduction

The first description of a subcellular structure termed "the nucleus" can be found in a paper first delivered to the Linnean Society of London in November of 1831. In his paper, the Scottish botanist Robert Brown – perhaps better known for his discovery of Brownian motion – concludes his observations on Orchids as follows.

I. Meier
Department of Plant Cellular and Molecular Biology and Plant Biotechnology Center,
Ohio State University, 244 Rightmire Hall, 1060 Carmack Road, Columbus, OH 43214, USA
e-mail: meier.56@osu.edu

Plant Cell Monogr, doi:10.1007/7089_2008_24

He mentions that "in each cell of the epidermis of a great part of this family, especially of those with membranaceaus leaves, a single circular areola, generally somewhat more opake than the membrane of the cell, is observable... This areola, or nucleus of the cell as perhaps it might be termed, is not confined to the epidermis, being also found not only in the pubescence of the surface... The nucleus of the cell is not confined to Orchideae, but is equally manifest in many other Monocotyledonous families; and I have found it, hitherto however in very few cases, in the epidermis of Dicotyledonous plants" (Brown 1833).

To the satisfaction of those currently interested in plant nuclear organization, the nucleus was thus indeed first described in plants. Nevertheless, it cannot be ignored that since John Brown's first observations, the field of nuclear biology has been mostly driven by non-plant model systems. More recently however, plant nuclear biology has begun to catch up and an active group of researchers is now addressing the connection between structure and function and the integration of mechanistic and organizational aspects of nuclear biology in higher plants. The chapters in this volume give an overview of our current understanding and highlight exciting new insights, emerging areas, as well as open questions.

2 Establishing Boundaries: Separation of Cytoplasm and Nucleoplasm

The first proposal of a membrane barrier that separates the nucleoplasm from the cytoplasm was published in 1913 based on micromanipulation studies of animal and plant cells (Kite 1913). Later, electron micrographs revealed that the nuclear envelope (NE) consists of two parallel membranes. At the same time, nuclear pores were identified that penetrate both membranes and it was demonstrated that such pores exist in a number of different tissues and cell types of rat, thus establishing nuclear pores as a general feature of the NE (Watson 1955).

Since those early studies, much has been learned about the composition and function of the nuclear envelope, while at the same time a number of important questions still await answering. Important are the recent advances in cataloging the proteins associated with the three types of NE membranes, the outer nuclear membrane (ONM), nuclear pore membrane, and inner nuclear membrane (INM). Proteomic approaches have identified about 30 different nucleoporins in the yeast and mammalian nuclear pore complex (NPC) and similarly, the number of known ONM and INM proteins has increased dramatically over the past few years (Schirmer and Gerace 2005). In all three cases, a concept emerges of a subgroup of proteins directly associated with the membranes, such as the integral membrane proteins emerin at the INM, gp210 at the nuclear pore, or Nesprin at the ONM. Other proteins assemble based on protein–protein interactions, with an exciting new concept being the crosstalk between ONM and INM complexes through protein–protein interactions in the NE lumen (Crisp et al. 2006). All three subclasses of NE proteins in turn interact with proteins beyond the nuclear periphery.

NPC proteins are contacting chromatin-associated proteins and the complexes involved in mRNA processing and export. INM proteins are in touch with chromatin as well, while ONM proteins contact elements of the cytoplasmic cytoskeletons (Crisp et al. 2006; Starr and Han 2003; Stewart et al. 2007).

While progress in understanding the composition and function of all three parts of the NE appears to have entered the exponential growth phase in yeast and animal model systems, much remains to be learned about the equivalent structures in plants. In this volume, Evans et al. discuss our current understanding of the functional composition of the plant NE. While the structures of the NE and NPC appear similar in plants and other eukaryotes, knowledge cannot simply be transferred from the non-plant model systems to plants. This is clear already from the fact that plants (like yeast) do not possess recognizable orthologs of the proteins that constitute the animal nuclear lamina, a protein meshwork underlying the INM. We can only begin to speculate what this implies for plant nuclei in terms of the emerging functions of the nuclear lamina for chromatin association with the NE, physical stability of the nuclear envelope, and the cellular functions disrupted in the laminopathies (Capell and Collins 2006). Is this entire functional complex missing from plants, or is the analogous function simply still unrecognized, because the individual proteins are structurally not similar to their animal counterparts? An exciting entry point into this question is the recent finding that depleting members of a somewhat "lamin-like" group of plant proteins (NMCPs) leads to a stunning reduction in nuclear size in *Arabidopsis* (Dittmer et al. 2007). Moreno Díaz de la Espina in this volume discusses this and other evidence that suggests that a "lamina-like" structure might exist in plants, but must be rather different in its molecular composition from its animal counterpart.

3 Nucleocytoplasmic Transport: The Ins and Outs of It

One of the great functional differences between eukaryotes and prokaryotes is the spatial separation of transcription and translation in eukaryotic cells. This alone implies a large amount of traffic between the two compartments in which these processes reside, the nucleus and the cytoplasm. As far as we know, all macromolecular traffic occurs through the nuclear pore. Much traffic of small molecules might use this passage too, however comparatively little is known about small-molecule traffic into and out of the nucleus, a field with great potential as we begin to learn more about the subcellular compartmentalization of regulatory small molecules and ions.

While a number of the core components of nuclear import and export appear to be conserved in plants, some differences are also apparent. Structurally, the plant nuclear pore closely resembles the animal or yeast nuclear pore, including the well-known eightfold symmetry (Roberts and Northcote 1970). Rose describes in this volume our current understanding of shared topics as well as unique aspects of the plant nuclear pore composition, protein nuclear import, and its accessory proteins such as the components of the Ran cycle. In another chapter, Merkle discusses our knowledge of plant nuclear export of proteins and RNA.

Recently, nuclear pore proteins and members of the import and export machinery have received significant attention from the plant community, when they surfaced in mutant screens of developmental and environment-response pathways. The respective pathways include the regulation of flowering time, hormone and stress responses, as well as plant–microbe interactions (reviewed in Xu and Meier 2008). While the precise role of nucleoporins and import or export components is currently not known, the emerging picture is that in all cases important factors of the pathways must become limiting either in the nucleus or in the cytoplasm. These new mutants might help to address a long-standing question in the nucleocytoplasmic trafficking field, namely whether there are components (such as individual members of the importin alpha family) that are specific for the transport of specific molecules or molecule subclasses. Stunningly, it has been recently demonstrated that substituting the prevalent importin alpha in a pluripotent human cell type with the one dominating in differentiated neural cells was sufficient to induce the expression of neuron-specific transcription factors (Yasuhara et al. 2007). In further approaching this question now in plants, it will be important to distinguish between true specificity and threshold effects, for example that some pathways might be more sensitive than others to a generic reduction in nucleocytoplasmic trafficking.

4   Intranuclear Architecture: Where to Hang Your HAT

Not too long ago, it was still generally assumed that chromatin-associated processes such as transcription and replication occurred randomly within the nucleus, with no apparent spatial organization. The past decade has fundamentally changed this view with technical advances from chromatin painting to epigenomics and we begin to see that nuclear processes are indeed architecturally organized within the highly complex nuclear landscape.

These advances also allow reinvestigation of the old concept of the "nuclear matrix" or "nucleoskeleton." Moreno Díaz de la Espina in this volumes reviews past evidence for such a structural framework of the nucleus and compares and contrasts it with the more recent concepts of subnuclear organization. While the "nuclear matrix" has been a topic of hot debate, the effects of the complementary DNA elements (Matrix Attachment Regions, MARs) have been established beyond doubt. Interestingly, this field has been kept very active from the onset by plant groups, originally driven by the problem of large variation in gene expression levels in individual transgenic plant lines, undesired for the development of robust transgenic crop plants (Allen et al. 2000).

As pointed out by Allen in this volume, much research in plants has therefore focused on how to best reduce variation in gene expression by adding well-defined MARs to transformation vectors. In contrast, in the yeast and animal fields more emphasis has been placed on understanding the role MARs play in higher-order chromatin structure and nuclear architecture. The questions for plant researchers are now obvious and include whether MARs are involved in creating higher-order chromatin structures in the plant nucleus, the relationship between MARs and histone-modification sites, the possible association of MARs with the nuclear pore, and the connection between MARs and transcriptional and post-transcriptional gene silencing. With much knowledge accumulated about individual plant MARs, the time has come to move this field now to the next level and use MARs as a tool to probe into the different aspects of plant chromatin organization. Most relevant for this paradigm shift will be the recent approaches to map MARs genome wide in the model plants *Arabidopsis* and rice and to correlate and functionally connect their distribution with other aspects of epigenomics such as histone modification and DNA methylation patterns, as discussed by Allen in this volume.

One of the most exciting discoveries of the past decade was that of the histone code, as the sum of all covalent modifications of the histone tails on the surface of the nucleosomes is now generally called. While it has been known since the 1970s that the core histones can be covalently modified, the function of these modifications and their relevance for epigenetic regulation have only become evident over the past 10 years (Turner 2000). Histone tails are modified by histone acetyl transferases (HATs) and histone methyl transferases, and are phosphorylated or ubiquitinated. Histone deacetylases (HDACs) and the respective enzymes for the other modifications can remove the individual tags. Individual modifications increase or decrease the likelihood of second nearby modifications, and different modifications of the same residue exclude each other. The resulting complex code is "read" by chromatin-binding proteins with affinity for the modified surfaces of the nucleosomes, such as bromodomain or chromodomain proteins. These proteins then lead to a series of events that might result in active, euchromatic or inactive, heterochromatic chromatin domains and can target DNA methyltransferases to chromatin. DNA methylation is thus tightly connected to histone modifications and methylated DNA can in turn recruit histone-modifying enzymes.

Reviewing the current literature in this field as well as his own body of work, Fransz in this volume concludes that plants and animals do not show large differences in the underlying epigenetic modules. Variations, however, exist in the combination of protein modules in the proteins that read the histone code, which might in turn reflect different interpretations of the epigenetic code.

Much has been learned recently about chromosome territories and chromocenters in plants, and in this field, comparison between species with small and large genomes will likely further our understanding of the role of the vast intergenic regions of some plant species. A major subnuclear domain is the nucleolus and exciting work is forthcoming about its chromatin organization. Together with recent proteomic studies of purified nucleoli (Pendle et al. 2005), we begin to see that the nucleolus might be far more than a ribosome biogenesis machine. Both for plants and animals alike, this field will likely surprise us with exciting new functions of this long-known sub-nuclear compartment.

5   Strategic Invaders: How to Conquer a Plant Nucleus

The use of the soil bacterium *Agrobacterium tumefaciens* for stabile genetic transformation of plants was first established in the 1980s. It takes advantage of a preexisting process that allows *Agrobacterium* to insert a fragment of its own DNA (the T-DNA) into the plant genome and re-program the respective plant cells to grow into undifferentiated tissue and provide nutrients for the bacteria (crown gall disease). It has by now become such commonplace procedure in most plant labs that only the rare and exceptional graduate student will still reflect on its molecular process. Those who do will soon notice that while much is known about the nature of the T-DNA that is transferred from *Agrobacterium* to the plant, the actual process of its trafficking from the plant cytoplasm to the nucleus and its subsequent integration into the plant genome are an area of active, exciting research with a significant number of unknowns.

As discussed by Danfi-Yelin et al. in this volume, we have only appreciated in the past few years how complex the process of integrating the T-DNA into the plant nucleus really is. We begin to understand to what degree the process relies on host factors supplied by the plant, as well as the plant's genome structure and DNA-repair machinery. For example, the nuclear import of the protein–DNA complex harboring the T-DNA (the T-complex) relies on both the *Agrobacterium* proteins VirE2 and VirD2 and on the host protein VIP1, which is taken hostage by the T-complex to provide part of the nuclear import signal required for traffic through the nuclear pore.

Once inside the nucleus, more host proteins are recruited to allow integration into the plant genome, among them a variant of plant histone H2A and a number of proteins involved in DNA repair. While there are currently still competing models for the exact T-DNA integration process, it is already apparent that the process is complex, and that its detailed study will likely teach us much not only about T-DNA integration but also about the host mechanisms involved in it.

6   Outlook: The Road Ahead

While much progress has been made during the past decade to further our understanding of plant nuclear functional organization, several areas have still been barely addressed or are still understudied. A long-standing open question in the field is the molecular nature of the plant equivalent of the animal centrosome. It is known that the outer surface of the nuclear envelope has microtubule nucleating activity and it has been proposed that it has centrosome-like activity (Vaughn and Harper 1998). However, how this feature exactly relates to plant spindle organization and what the molecular players are is still unclear. Similarly, while the dynamic changes of the metazoan nuclear envelope during cell division is now an active area of research (Anderson and Hetzer 2008), a comparable understanding of the processes that govern the orchestration of open mitosis in plants is an exciting field yet to emerge. Related to the fate and role of the nuclear envelope during mitosis is the function of nuclear pore proteins in the process. In animals, a number of nuclear pore components have been shown to traffic to the kinetochores and to be involved in the kinetochore checkpoint (Arnaoutov and Dasso 2003). It will be important to perform similar studies for the plant proteins to establish how similar or different their function is during plant mitosis. During interphase, the small ubiquitin-like protein modifier SUMO is gaining increasing importance as a protein tag likely involved in subnuclear localization and possibly in regulated nucleocytoplasmic trafficking. Understanding its precise function in a variety of important nuclear processes is currently a wide-open and challenging field in any system and therefore an exiting area for the early and timely involvement of plant biologists.

References

Allen GC, Spiker S, Thompson WF (2000) Use of matrix attachment regions (MARs) to minimize transgene silencing. Plant Mol Biol 43:361–376

Anderson DJ, Hetzer MW (2008) The life cycle of the metazoan nuclear envelope. Curr Opin Cell Biol 20:386–392

Arnaoutov A, Dasso M (2003) The Ran GTPase regulates kinetochore function. Dev Cell 5:99–111

Brown R (1833) On the organs and mode of fecundation in Orchideae and Asclepiadeae. Trans Linnean Soc London 16:685–745

Capell BC, Collins FS (2006) Human laminopathies: nuclei gone genetically awry. Nat Rev Genet 7:940–952

Crisp M, Liu Q, Roux K, Rattner JB, Shanahan C, Burke B, Stahl PD, Hodzic D (2006) Coupling of the nucleus and cytoplasm: role of the LINC complex. J Cell Biol 172:41–53

Dittmer TA, Stacey NJ, Sugimoto-Shirasu K, Richards EJ (2007) LITTLE NUCLEI genes affecting nuclear morphology in *Arabidopsis thaliana*. Plant Cell 19:2793–2803

Kite G (1913) The relative permeability of the surface and interior portions of the cytoplasm of animal and plant cells. Biol Bull 25:1–7

Pendle AF, Clark GP, Boon R, Lewandowska D, Lam YW, Andersen J, Mann M, Lamond AI, Brown JW, Shaw PJ (2005) Proteomic analysis of the Arabidopsis nucleolus suggests novel nucleolar functions. Mol Biol Cell 16:260–269

Roberts K, Northcote DH (1970) Structure of the nuclear pore in higher plants. Nature 228:385–386

Schirmer EC, Gerace L (2005) The nuclear membrane proteome: extending the envelope. Trends Biochem Sci 30:551–558

Starr DA, Han M (2003) ANChors away: an actin based mechanism of nuclear positioning. J Cell Sci 116:211–216

Stewart CL, Roux KJ, Burke B (2007) Blurring the boundary: the nuclear envelope extends its reach. Science 318:1408–1412

Turner BM (2000) Histone acetylation and an epigenetic code. Bioessays 22:836–845

Vaughn KC, Harper JD (1998) Microtubule-organizing centers and nucleating sites in land plants. Int Rev Cytol 181:75–149

Watson ML (1955) The nuclear envelope; its structure and relation to cytoplasmic membranes. J Biophys Biochem Cytol 1:257–270

Xu XM, Meier I (2008) The nuclear pore comes to the fore. Trends Plant Sci 13:20–27

Yasuhara N, Shibazaki N, Tanaka S, Nagai M, Kamikawa Y, Oe S, Asally M, Kamachi Y, Kondoh H, Yoneda Y (2007) Triggering neural differentiation of ES cells by subtype switching of importin-alpha. Nat Cell Biol 9:72–79

The Plant Nuclear Envelope

David E. Evans (✉), **Sarah L. Irons Katja Graumann, and John Runions**

Abstract The nuclear envelope is an important but poorly studied dynamic membrane system in plants. In particular, surprisingly little is known about the proteins of the higher plant nuclear envelope and their interactions. While structurally similar to the nuclear envelope of other kingdoms, unique properties suggest significant differences. For instance, plants lack sequence homologues of the lamins and instead of centrosomes the entire nuclear envelope surface acts as a microtubule-organising centre. This chapter reviews the structure of the nuclear envelope in relation to its protein domains, namely the inner and outer membrane, and the pore domain. Recent advances in the characterisation of novel proteins from these domains are presented. In addition, new insights into mechanisms for the targeting and retention of nuclear envelope proteins are discussed. The nuclear envelope is of importance in cell signalling and evidence for physical nucleo-cytoskeletal linkage and for the nucleoplasm and periplasm as calcium signalling pools are considered. Finally, the behaviour of inner nuclear membrane proteins during the breakdown and reformation of the nuclear envelope in mitosis is discussed.

1 Introduction

The nuclear envelope (NE) is a complex structure separating cytoplasm from nucleoplasm and is a defining characteristic of eukaryotic cells. In electron micrographs it appears as a two-membrane system with the lumen perforated by nuclear pores, but static images do not convey its complexity and dynamic interactions with the nucleoplasm and nucleoskeleton and with the cytoplasm and cytoskeleton.

D.E. Evans
School of Life Sciences, Oxford Brookes University, Headington Campus,
Oxford, OX3 0BP, UK
e-mail: deevans@brookes.ac.uk

Plant Cell Monogr, doi:10.1007/7089_2008_22

Such interactions require an array of proteins with complex binding interactions as well as mechanisms for transport through the pores, traffic and targeting within the membrane, breakdown and reformation during cell division and for positioning the nucleus within the cell.

Detailed study of the proteins of the plant NE has only recently been undertaken and this chapter will consider advances in knowledge and understanding of its protein constituents and their binding partners largely achieved over the last five years. Large gaps in knowledge remain; in particular key "missing" links whose presence is suggested by current data. This means that it remains an important and under-researched area. There are a number of unique features of the plant NE; hence knowledge from other organisms cannot be generalized to plants. In particular, plants lack structures analogous to centrosomes and the entire outer NE serves as a microtubule-organising centre (MTOC) (Shimamura et al. 2004). The plant NE is associated with the cytoskeleton and nucleoskeleton, but plant nuclei lack sequence homologues of the nuclear lamins (see Morena Diaz de la Espina 2008; and Brandizzi et al. 2004). Thus, understanding the interactions of the plant NE with elements of both cytoskeleton and nucleoskeleton is a particularly important area.

2 Nuclear Envelope Domains

The nuclear envelope is a double membrane surrounding the nuclear material. In it, three separate but linked domains have been identified; the outer nuclear membrane (ONM), the inner nuclear membrane (INM) and, the membrane connection between them, the pore membrane (POM) within the nuclear pore complex (NPC). In addition, invaginations collectively called the nucleoplasmic reticulum (see Prunuske and Ullman 2006) or transnuclear strands (Fricker et al. 1997; Collings et al. 2000) penetrate the nucleoplasm and greatly increase the surface area of the INM and its proximity to chromatin in certain cell types. The nuclear pore complexes are involved in maintaining the individual composition of the three membrane domains as proteins can only diffuse into the INM through them (Mattaj 2004). There is an increasing body of evidence to suggest that INM proteins require a nuclear targeting sequence recognised for passage through the pore (Lusk et al. 2007). In addition, protein binding also strongly influences the constituents of each of the domains with protein–protein interactions acting to anchor and thereby enrich proteins within them.

3 The Outer Nuclear Membrane

The ONM is in close association with perinuclear endoplasmic reticulum (PNER) and is linked to it through junctional regions that allow traffic of PNER proteins into the ONM and vice versa (Staehelin 1997). The ONM is also frequently decorated with ribosomes and functions in protein synthesis (Gerace and Burke 1988). The ONM appears to share most of its protein constituents with the PNER, but the two proteomes, while over-lapping, are not identical and proteins predominantly located at the ONM have been identified (Schirmer and Gerace 2005). The abundance of proteins also differs between the two domains, though their proximity and close connection makes separating them for biochemical quantification difficult if not impossible. It is suggested that the presence of physical constrictions of 25–30 nm in diameter in the junctional regions restrict protein movement from endoplasmic reticulum (ER) to ONM (Craig and Staehelin 1988; Staehelin 1997).

The proteins of the ONM play a vital role in connecting the nucleus to the cytoskeleton, for nuclear positioning, maintaining the shape of the nucleus, signal-ling and cell division. The bridge between the INM and ONM described in animal systems involves proteins of the LINC (LInker of Nucleoskeleton and Cytoskeleton) complex (Crisp et al. 2006). In the LINC complex, proteins possessing a transmem-brane domain spanning the ONM interact via a conserved binding domain with partners that span the INM and interact with nucleoskeletal proteins. These interac-tions prevent the ONM protein from moving back into the PNER thereby retaining it in the ONM. The LINC complex is considered in detail later in the chapter.

While definite protein constituents of the plant ONM still remain elusive, one possible candidate is DMI1 (doesn't make infection 1). This putative ion channel of *Medicago truncatula* was found in a screen for proteins essential in establishing symbiotic relationships between plants and nitrogen-fixing bacteria (Riely et al. 2006). It was shown to be involved in generating perinuclear calcium oscillations in response to Nod factor signalling. Nod factor is released by nitrogen-fixing bac-teria and the corresponding cellular ion fluxes, calcium spikes, lead to the activation of early nodulation genes, which result in root hair nodulation and thus the symbi-otic interaction between the plant root and the bacteria (Riely et al. 2006; Peiter et al. 2007). Although DMI1 is an ion channel it is not thought that Ca^{2+} from the periplasmic space uses this channel to efflux into the cytoplasm. Instead, DMI1 has a C-terminal RCK (Regulator of the Conductance of K^+) domain, which when deleted abolishes calcium spikes and interferes with nodule formation. A Green Fluorescent Protein (GFP) fusion to DMI1 was used to localise the protein to the NE. Intriguingly, *Lotus japonicus* homologues CASTOR and POLLUX are localised to plastid membranes in onion and pea roots (Riely et al. 2006; Peiter et al. 2007).

In addition to ONM intrinsic proteins there are also a number of peripheral and soluble proteins that associate with the cytoplasmic face of the outer nuclear membrane. For instance, RanGAP (see Rose 2008) has been shown to decorate the plant ONM. While it has been long known that the entire surface of the plant NE has microtubule (MT) nucleating activity (Stoppin et al. 1996) only recently two proteins were identified to be essential for this function. While animal cells have one MTOC that nucleates MTs, plant cells have so far been shown to have at least three distinct regions with such a function – the NE surface, the cortex underlying the plasma membrane and branching points of pre-existing MTs (Starr and Han 2003). Intrinsic to eukaryotic MT nucleation sites is the gamma-tubulin ring complex (gamma-TuRC), which consists of five gamma-tubulin complex proteins (GCP) and gamma-tubulin itself (Fava et al. 1999; Murphy et al. 2001). The plant proteins AtGCP2 and AtGCP3 are homologues of *Drosophila* and yeast GCP2 and 3, and have been shown to form a soluble complex with gamma-tubulin that associates with the plant ONM (Seltzer et al. 2007). Intriguingly, both have NE targeting domains that are thought to target the gamma-TuRC to the NE, which is then retained there by associating with an as yet unknown ONM intrinsic protein (Seltzer et al. 2007).

4 The Pore Membrane

The proteins of the nuclear pore and the mechanisms of transport through it have been discussed in detail elsewhere in this volume (Rose 2008). The discussion here will therefore be limited to the proteins and role of the pore membrane. The pore membrane is the only membrane connection between INM and ONM and therefore all proteins destined for the INM must pass through it. It is also the point of anchorage of the proteins of the pore and is important in nuclear pore (and therefore NE) formation.

A number of pore domain proteins have been identified in metazoans and yeast. In a comparative genomic study, Mans et al. (2004) identified five pore membrane proteins (POMS); Pom34, Pom152, Pom 210, Ndc1p and Pom121. They observed that Pom34 and Pom152 were restricted to fungi, Pom121 was unique to vertebrates and only Pom210 and Ndc1 were present in animals, fungi and plants.

Pom34 and Pom121 contain two transmembrane domains that span the pore membrane. Both the N-terminus and the C-terminus of Pom34 reach into the pore and are involved in the molecular organisation of the pore complex (Miao et al. 2006). On the other hand, Pom121 is required for NE formation. Depletion of Pom121-containing membrane vesicles does not affect vesicle binding to chromatin but fusion to form a closed NE does not occur (Antonin et al. 2005) Thus, it appears that Pom121 links formation of the nuclear pore complex with the reassembly of the NE.

Glycoprotein 210 (Pom210/gp210) is a major component of the nuclear pore complex (Courvalin et al. 1990). It contains a large perinuclear N-terminal domain that comprises 95% of its mass and contains an Ig-fold domain, which allows interactions between gp210 and other NE components (Greber et al. 1990; Mans et al. 2004). A single transmembrane (TM) segment anchors the protein into the pore membrane and is followed by a short C-terminal cytoplasmic tail. The TM domain is thought to contain a nuclear targeting signal as it is essential and sufficient for correct localisation of gp210 (Wozniak and Blobel 1992). For it to properly function, gp210 forms homodimers (Mans et al. 2004). Homologues of gp210 are present in *Arabidopsis* and are predicted to be structurally similar to the mammalian and *Caenorhabditis elegans* proteins, with a C-terminal transmembrane region (Cohen et al. 2001).

Ndc-1, a 74-kDa protein with six transmembrane domains, has been identified in the pore membranes of animals and yeast (Winey et al. 1993; Chial et al. 1998).

Stavru et al. (2006) demonstrated that it is also present in spindle pole bodies of *S. cerevisiae*. Ndc1 interacts with soluble Nups such as Nup53 and is therefore essential for anchoring the nuclear pore complex (NPC) to the membrane (Mansfeld et al. 2006). It also plays a role in NPC assembly, in particular in correct assembly of Nups with FG repeats that are implicated in cargo trafficking through the pore (Stavru et al. 2006). On the basis of homology, Mans et al. (2004) suggest that an Ndc1 homologue is present in plants; its localization and interactions, however, are as yet unknown.

Apart from gp210 and Ndc1, plants also possess a family of unique pore membrane proteins. Two of the three WIPs (WPP interacting protein) localise to the ONM/pore membrane in *Arabidopsis* and homologues have also been found in other plant species (Xu et al. 2007). The WPP domain and its importance in RanGAP targeting has been discussed elsewhere (see Rose 2008). Similar to gp210, WIPs1-3 contain a TM domain and nuclear targeting signal that ensure correct localisation to the ONM and pore membrane. A coiled-coil domain at the cytoplasmic side is thought to mediate both dimerisation and association with the WPP domain of RanGAP. Whether the WIPs are also involved in the assembly, anchorage and maintenance of NPC is as yet unsolved. While all three WIPs seem to functionally overlap, a triple knockout abolishes NE anchorage of RanGAP. Curiously, this only occurs in *Arabidopsis* root tips and does not affect the development and growth of the plant. Xu et al. (2007) therefore speculate that RanGAP may be dispensable in root tip nuclear transport and suggest that the members of the WIP family are fairly newly evolved plant ONM/pore membrane proteins. It certainly demonstrates that RanGAP anchorage to the NE can be cell-type specific, a phenomenon that has not been shown in animals so far (Xu et al. 2007).

Thus, there appears to be a divergence between kingdoms in the protein composition of the pore membrane and therefore the anchorage of pore proteins. In plants, only Pom210 and Ndc1 have thus far been identified as components of this ring, though others may yet be discovered. This places one protein with six transmembrane domains and a second single pass protein with a very large luminal domain into this region; it is suggested that they are arrayed in a ring beneath the nucleoporins. As referred to above, all inner nuclear membrane proteins have to traffic through this domain to either enter or leave this membrane and mechanisms for this traffic are therefore of particular interest.

5   The Inner Nuclear Membrane

A number of proteins of the inner nuclear membrane have been characterised in animal cells. Excitement in this field stems, at least in part from the growing number of human diseases such as muscular dystrophies, lipoatrophy, skeletal defects and epilepsy stemming from INM protein mutations (e.g. Wilkie and Schirmer 2006). The variety and function of INM proteins is significant. For instance, LEM domain proteins (Lamin Associated Proteins (LAPs), Emerin, Man) interact with BAF (barrier to autointegration factor) and are involved in gene silencing (Gruenbaum et al. 2005). In addition, emerin associates nuclear actin to the INM and Man plays an antagonising role in the TGF-beta (transforming growth factor beta) signalling cascade (Gruenbaum et al. 2005; Bengtsson 2007). ASI1-3 (amino acid sensor independent) are also signal transducing proteins that prevent inappropriate expression of amino acid permeases (Zargari et al. 2007). Another well-studied mammalian INM protein is the lamin B receptor (LBR). It has its own multimeric protein complex and apart from a sterol reductase activity binds to lamin B, chromatin and is involved in RNA splicing (Chu et al. 1998; Nikolakaki et al. 1997). Other INM proteins connect with the nucleoskeleton, in particular the lamina, and SUN (Sad1/Unc84) proteins even link the nucleoskeleton with the cytoskeleton via the LINC complex (Crisp et al. 2006; see below). Overall, the INM and its proteins are implicated in nucleic acid metabolism, signal transduction, NPC spacing and the tethering of nuclear matrix and chromatin. Most of these functions also involve the lamina as the lamina is tightly associated with the animal INM and is essential for its integrity (Gruenbaum et al. 2005). As no plant homologues of lamins seem to exist (Irons et al. 2003; Rose et al. 2004; Graumann et al. 2007; also see Moreno Diaz de la Espina 2008) it is not surprising that many of the well-characterised animal INM proteins do not appear to have plant homologues either – apart from two notable exceptions (Table 1).

Firstly, two *Arabidopsis* homologues of the yeast SUN protein Sad1 were identified by Van Damme et al. (2004) in a screen for proteins implicated in plant cytokinesis and phragmoplast formation. GFP fusions of AtSad1a and AtSad1b showed them to be localised to the NE in interphase and a putative bipartite nuclear localisation signal suggests that both may be present in the INM portion of the NE (van Damme et al. 2004; Graumann K, unpublished observations). Van Damme et al. (2004) also used the GFP fusions to examine for the first time the fate of plant NE proteins in mitosis. Prior to NE breakdown AtSad1a was observed in dots associated with the nuclear rim and close to the plasma membrane. After NE breakdown the construct surrounded the spindle and phragmoplast. Interestingly in metaphase the proteins had accumulated in a few bright dots at both ends of the spindle, which resembled yeast Sad1 bodies. From this van Damme et al. (2004) argue that AtSad1a and AtSad1b may be involved in MT nucleation during mitosis as well as other functions that the SPB and MTOC fulfil in dividing yeast and animal cells respectively, such as telomere clustering. Our findings that AtSad1a and AtSad1b contain the highly conserved SUN domain (see LINC complex) would support such a hypothesis (Graumann and Evans, unpublished results).

The second family of plant INM proteins is in fact not membrane intrinsic but strongly associates with the NE very much like the earlier described components of the Ran cycle. The Aurora kinase family members regulate mitotic processes and two of the three *Arabidopsis* homologues have been found to localise to the NE (Kawabe et al. 2005). GFP fusions of these were followed throughout mitosis and were found to be present at the mitotic spindle, centromeres and the cell plate (Demidov et al. 2005; Kawabe et al. 2005). As serine-threonine kinases they have

Table 1 Proteins of the nuclear envelope and their putative plant homologues

Protein	Organisation	Function	Plant homologue	Reference
LAP1 and LAP2 (lamin associated proteins)	LAP1A/B/C isoforms LAP2β/γ/δ/ε/ξ isoforms All have LEM (Lamin, Emerin, Man) domain	LEM interacts with BAF (barrier to auto integration factor), which cross links DNA with LEM domain proteins DNA replication and expression NE expansion after mitosis	None LEM domain restricted to metazoa	Foisner (2001); Gruenbaum et al. (2005)
Emerin	LEM domain	Interacts with nuclear f-actin, lamins and BAF Transcriptional regulator, death promoting repressor	None	Gruenbaum et al. (2005)
Man1	LEM domain	Binds BAF and Smads and transcriptional regulators Antagonises TGF β signalling	None	Bengtsson (2007)
LBR (lamin B receptor)	Eight TM with sterol reductase activity N′ terminal binds lamin and chromatin	N-terminus associates with b-type lamins, chromatin and other components of its own multimeric complex and is involved in RNA splicing and chromatin organisation	None	Chu et al. (1998)
Nurim (nuclear rim)	Six TM	Tightly associates with nuclear matrix Is hypothesised to have enzymatic action on lamins	None	Hofemeister and O'Hare (2005)
Nesprins	Nesprins-1/2/3 Predominately ONM proteins but isoforms present in INM	In ONM part of the LINC complex In INM interacts with emerin and lamins Speculated involvement in lamina organisation	None	Wilhelmsen et al. (2006)
RFBP (ring finger binding protein)	Ring finger binding domain Type IV p-type ATPase	Phospholipid transporting ATPase Sub nuclear trafficking of transcription factors with Ring domain	Four putative Arabidopsis homologues – ALA4/5/11/12	Mansharamani et al. (2001) Graumann K. (unpublished observation)[a]
SUN (Sad1/UNC-84 homology)	SUN1 and SUN2 C-terminal SUN domain	Interact with lamins on nucleoplasmic side and nesprin at periplasmic side Involved in nucleo-cytoskeletal bridging	Two putative Arabidopsis homologues AtSad1a and AtSad1b	Tzur et al. (2006b); Van Damme et al. (2004) Graumann et al. (unpublished results)
NCX – sodium calcium exchangers		Facilitate transport of Ca^{2+} from nucleoplasm to perinuclear space Interactions with GM1 potentiate exchange	None	Ledeen and Wu (2007)

[a]BLAST search of rabbit RFBP identified hits in Medicago truncatula, Oryza sativa and Arabidopsis thaliana; four of the six Arabidopsis homologues have putative bipartite nuclear localisation signals (Motif Scan) and one (ALA5) is predicted to be nuclear (WoLF PSORT)

been shown to phosphorylate histone H3 and in view of their cellular locations they have been implicated in chromosome segregation and cytokinesis (Demidov et al. 2005; Kawabe et al. 2005).

So far AtSad1a, AtSad1b and the Aurora kinases are the only identified plant INM components. Clearly with the plant INM implied in a variety of important nuclear and cellular functions there is great potential for discovering more of its components. A lack of lamins in plants suggests that other lamin-like proteins functionally replace lamins in plants (see Moreno Diaz de la Espina 2008) and that the plant NE proteome may vary considerably from its animal and yeast counterpart (Meier 2001; Brandizzi et al. 2004).

6   Targeting and Retention of INM Proteins in Plants: Studies with the Lamin B Receptor

Bi-directional transport of RNAs and soluble proteins through the NPC are explained in detail in the chapters by Merkle (2008) and Rose (2008). This highly regulated process enables the cell to control the type and amount of molecules that enter and exit the nucleus and therefore DNA and RNA metabolism as well as translation and protein synthesis. This ultimately affects all aspects of nuclear and cellular function. It seems rational to argue that transport of INM proteins would also be regulated as they play crucial roles in chromatin organisation (Lusk et al. 2007). A basic model for correct INM protein localisation is the targeting-retention model (Mattaj 2004). It suggests that INM proteins are co-translationally inserted into the ER membrane and from there diffuse through the ER-ONM-pore membrane continuum into the INM, where they are tethered by protein interactions and thus accumulate. Other membrane intrinsic proteins such as ER and ONM proteins would diffuse out of the INM and are not retained there due to a lack of interaction partners. This model has since been expanded by growing research that points to a more controlled approach of INM protein targeting similar to that of soluble proteins (Lusk et al. 2007). It is thought that INM protein targeting already commences during co-translational insertion into the ER membrane. Both mammalian INM proteins LBR and nurim as well as viral INM proteins were shown to have a sorting signal in their first TM domain, which is recognised by the components of the translocation channel and results in active targeting of the INM proteins to the NE mediated by importin beta (Saksena et al. 2004; Saksena et al. 2006). How the INM proteins pass the pore membrane is still not clear but it has been demonstrated that the process requires energy and results in restructuring of the NPC (Ohba et al. 2004; King et al. 2006; Lusk et al. 2007). Lusk et al. (2007) suggest a bimodal system in which membrane proteins with a cytoplasmic/nucleoplasmic domain smaller then 25 kDa may diffuse through the pore but proteins with larger domains are actively transported. Nuclear localisation signals (NLS) previously found in soluble proteins have also been identified in INM proteins such as LBR and SUN2 and it is thought that NLS and karyopherin-mediated transport is advantageous and even essential for INM protein import (Lusk et al. 2007). In addition to karyopherins, interactions with certain Nups might also be crucial. For instance, deleting Nup170 disrupts INM protein but not soluble protein import (Lusk et al. 2007). Once INM proteins arrive at their destined membrane, binding interactions with lamins, chromatin and other INM proteins are thought to retain them in the INM.

Using a truncated LBR fused to GFP the authors have also studied INM protein targeting in plants (Irons et al. 2003, Graumann et al. 2007). The full length LBR protein has eight transmembrane domains but the N-terminal nucleoplasmic domain and first transmembrane domain are sufficient for INM localisation in animal cells (Ellenberg et al. 1997). When this truncated mammalian LBR containing a bipartite NLS and single transmembrane domain is expressed in a plant system, it localises to the NE (Irons et al. 2003; Fig. 1). Our recent studies suggest that the mechanisms for targeting LBR to the INM are conserved between kingdoms and traffic through the plant pore membrane occurs in the same way as in mammalian cells (Graumann et al. 2007). In addition, it has revealed that strong binding interactions in the INM are not necessary for retention. In mammalian cells the N-terminal domain of LBR binds to lamin B and chromatin, and thereby becomes immobilised at the INM. However, lamin B homologues do not exist in plants and FRAP analysis showed that LBR-GFP is highly mobile in the plant INM (Graumann et al. 2007). The authors favour a model in which LBR is trapped in the INM as so far no nuclear export signal (NES) could be identified in its entire sequence. Building on the theory that INM proteins require NLS and karyopherins for import like soluble proteins it can be hypothesised that an NES and exportins may be involved in export of the protein. Instead, the INM protein remains in the INM and may eventually be degraded by a proteasome pathway (Graumann et al. 2007).

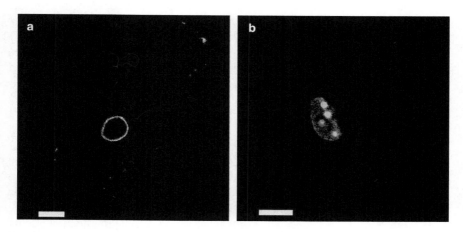

Fig. 1 LBR-GFP locates to the NE when transiently expressed in tobacco leaf epidermal cells (**a**); when a putative binding domain is mutated the protein is relocated to inclusions inside the nucleus (**b**). Scale bar = 10 μm. Graumann, K., unpublished

7   The Lumen and the LINC Complex

Although the NE separates the chromatin from the cytoplasm, the nucleus is not isolated from the rest of the cell. While NPCs provide aqueous channels for molecules to be exchanged between nucleus and cytosol, LINC complexes connect the cytoskeleton with the nucleoskeleton thereby presenting a further route of communication between the nucleus and the cell (Crisp et al. 2006). The composition and function of these bridging complexes have been subject to recent investigations in animal and yeast systems (Starr and Han 2003; Starr and Fischer 2005; Tomita and Cooper 2006; Tzur et al. 2006b; Wilhelmsen et al. 2006; Worman and Gundersen 2006), however very little is known about their existence in plants.

The key components of the LINC complex are ONM located proteins with a highly conserved KASH (Klarsicht/ANC-1/Syne Homology) domain and SUN domain proteins of the INM. The KASH domain in the perinuclear space then interacts with the SUN domain to create a bridge across the two membranes. This connection interlinks cytoplasmic microtubules, filamentous actin and intermediate filaments with nuclear lamins (Starr and Fischer 2005; Wilhelmsen et al. 2006; Crisp et al. 2006). To interact with the cytoskeletal components KASH domain proteins have large coiled-coil domains or spectrin repeat regions that can protrude up to 500 nm into the cytoplasm. Indeed, KASH domain proteins are huge molecules with the largest, human nesprin-1, being just over 1 MDa in size (Zhang et al. 2002; Padmakumar et al. 2004). Actin-binding KASH proteins, such as *C. elegans* ANC-1 protein, have two additional calponin domains at the N-terminus to directly associate with f-actin (filamentous actin) (Starr and Fischer 2005; Wilhelmsen et al. 2006; Worman and Gundersen 2006). *Schizosaccharomyces pombe* Kms1, *C. elegans* ZYG-12 and *Drosophila* Klarsicht proteins are examples of KASH domain proteins that associate with components of the MT cytoskeleton, in particular dynein (Mosley-Bishop et al. 1999; Starr and Fischer 2005; Wilhelmsen et al. 2006). Human nesprin-3 is the only KASH domain protein that has so far been shown to interact with intermediate filaments (Wilhelmsen et al. 2005). On the nuclear side of the LINC complex SUN domain proteins interact with lamins and chromatin. The transmembrane domain, nuclear localisation signals and lamin binding domains but not the highly conserved C-terminal SUN domain are necessary for INM localisation of the proteins. SUN domain proteins also have coiled-coils, which are thought to be involved in homo- and heterodimerisation so that multiple cytoskeletal and nuclear components can be cross-linked (Tzur et al. 2006a). Thus, coupling the nucleus to the cytoskeleton allows for controlled movement, positioning and anchorage of the nucleus inside the cell. The LINC complex also associates centrosomes and spindle pole bodies (SPB) to the NE in animal and yeast cells respectively (Starr and Han 2003; Starr and Fischer 2005) and is involved in centrosome duplication (Kemp et al. 2007). The physical connection between cytoskeleton and chromatin via KASH and SUN proteins has been shown to be necessary for clustering of telomeres and the formation and anchorage of the meiotic chromosome bouquet at the NE (Chikashige et al. 2006; Tomita and Cooper 2006; Schmitt et al. 2007).

Components of the LINC complex are also part of signalling pathways (Starr and Fischer 2005), for example during apoptosis (Tzur et al. 2006a). To date SUN and KASH domain homologues have been identified in the model organisms *Schizosaccharomyces pombe, Saccharomyces cerevisiae, Caenorhabditis elegans, Drosophila melanogaster, Mus musculus* and *Homo sapiens* (Starr and Han 2003; Starr and Fischer 2005; Tomita and Cooper 2006; Tzur et al. 2006a; Wilhelmsen et al. 2006; Worman and Gundersen 2006). However, very little is known about their existence in plants. Our recent investigations suggest that while KASH domain proteins may be less conserved in plants, SUN domain homologues exist (Graumann and Evans, unpublished observations). Their identification and characterisation should allow the first insights into the workings of a possible plant LINC complex.

8   The Nucleoplasmic Reticulum and Nucleoplasmic Signalling

The nucleoplasmic reticulum exists in many cells as invaginations of both inner and outer nuclear envelope, often traversing the nucleus. As their lumen is connected directly to the cytoplasm, they greatly increase the surface area of the nucleus and provide regions within the nucleus where nucleoplasm and cytoplasm are in close proximity (Fricker et al. 1997; Broers et al. 1999; Echevarria et al. 2003). They may be branched and contain nuclear pore complexes and are surrounded by the nuclear lamina. The nucleoplasmic reticulum is involved in nucleoplasmic signalling and has been shown to be a site of calcium release (Lui et al. 1998; Echevarria et al. 2003). It may also be involved in the organisation of chromatin via lamin-DNA contacts. Broers et al. (1999) and Lagace and Ridgway (2005) showed that proliferation of the nucleoplasmic reticulum in a number of animal cell types was stimulated by CTP: phosphocholine cytidyltransferase-a, a key enzyme in the CDP-choline pathway for phosphatidylcholine synthesis that shows fatty acid-stimulated localisation to the NE.

The NE and nucleoplasmic reticulum play a key role in intranuclear signalling involving calcium and the inositol trisphosphate pathway. The nucleoplasm acts as a separate signalling domain from adjacent cytoplasm with the NE achieving attenuation of signalling (Al-Mohanna et al. 1994). Calcium signals in the nucleus influence DNA repair and gene transcription and the nuclear envelope and nucleoplasmic reticulum as a calcium signalling pool (Marius et al. 2006); several Ca-responsive proteins have been identified in the nucleus including annexin, transcription factors, calmodulin, and calcium-dependent protein kinases and phosphatases (Bouche et al. 2005; Kalo et al. 2005).

In animals the nuclear envelope contains a number of key components of Ca-signalling pathways: inositol 1,4,5-trisphosphate receptors, ryanodine receptors and Ca-ATPase activity. Inositol 1,4,5-trisphosphate receptors initiate localised nuclear InsP3-mediated calcium signals and the Ca signal generated results in movement of nuclear protein kinase C to the nuclear envelope (Santella and Kyozuka

1997; Bootman et al. 2000; Marius et al. 2006). In plants, the nucleoplasm has been shown to act as a separate Ca-signalling domain (Pauly et al. 2000). Nuclear Ca responds to temperature and to mechanical stimulation and transient receptor potential (TRP)-like Ca channels have been suggested to be present (Xiong et al. 2004). Patch-clamping isolated red beet nuclei show that non-selective voltage-dependent Ca-channels are present (Grygorczyk and Grygorczyk 1998). Mathematical analysis of Ca transport in the nucleus suggests a model in which the lumen of the NE acts as a Ca-signalling pool into which Ca is accumulated by active transport and released by Ca channels (Briere et al. 2006). In this model, channels in the inner nuclear membrane release Ca into the nucleoplasm. A calcium transporter located in the inner membrane is predicted to pump Ca back into the lumen. The author's laboratory provided immunocytochemical evidence for a Ca-ATPase localised at the nuclear envelope (Downie et al. 1998) but it has not been possible to prove an INM location and the pump may be restricted to the ONM. Further evidence for such a pump at the ONM has been provided (Bunney et al. 2000).

9   The Nuclear Envelope and Mitosis

Plants in common with most eukaryotes, with the notable exception of fungi, undergo open cell division, whereby the NE breaks down at the onset of mitosis and reforms around the nuclei of the new daughter cells. Through the use of fluorescent protein fusions and live cell imaging it has been shown that some NE proteins relocate to the ER during NE disassembly in cultured animal (Ellenberg et al. 1997) and plant cells (Irons et al. 2003; Fig. 2). In the animal cells it was shown that the NE marker (LBR-EGFP) exhibits a significant change in mobility between interphase and mitosis, with LBR being predominantly immobile within the intact NE in contrast to showing a high diffusion rate within mitotic membranes (Ellenberg et al. 1997). The presence of NE proteins in the ER during mitosis has also been shown by immunolabelling of native proteins in mammalian cells (Yang et al. 1997). In plant cells the NE/ER mitosis story remains to be fully described as, despite the clear specific location of the protein to the NE, recent photobleaching data suggests that LBR-GFP shows a high level of mobility within the NE membranes at interphase, and apparently lacks the significant fraction of immobilised protein present in mammalian cells (Graumann et al. 2007).

The possible mechanisms responsible for NE breakdown (NEBD) have been well described in animal cells and to some extent in plants. In animal cells NEBD is the result of microtubule-dependent stretching of the nuclear lamina which causes deformation of the NE and changes in nuclear pore distribution (Beaudouin et al. 2002; Salina et al. 2002). Initial tearing of the NE occurs on the opposite side of the nucleus where the tensile forces are at their greatest; the tearing is possibly linked to the localised disassembly of nuclear pores creating a focal point for the membrane perforation. After the initial tearing of the nuclear membrane, structure of the NE is lost rapidly. In tobacco BY-2 cells co-expressing a microtubule marker (GFP-MBD)

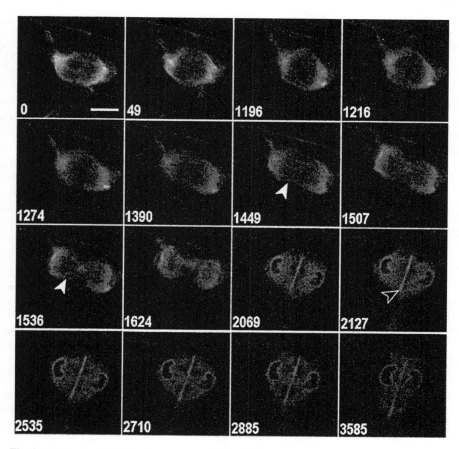

Fig. 2 Cell cycle in BY-2 cells expressing LBR-GFP as a nuclear envelope marker. Cell cycle progression is indicated in seconds as elapsed time from the start of the sequence. A metaphase BY-2 cell transformed with LBR-GFP (Irons et al. 2003) shows fluorescence distributed through-out the ER (time = 0–1,216 s). During the cycle progression, tubular membranous structures form through the MA (*arrowhead*, time 1,274–1,507 s). Subsequently, the ER membranes encircle the newly formed daughter nuclei (1,624 s) and the NE forms around each nucleus (2,069 s). The phragmoplast (*empty arrowhead*) forms between the nuclei and grows across the cell as more wall is assembled (2,069–3,585 s). Scale bar = 20 µm. Reproduced from Irons et al. (2003) with permission

and an ER/Golgi marker (Nag-DsRed) to highlight the NE it was observed that NEBD occurred before the disappearance of the pre-prophase band (PPB), a MT structure characteristic of plant cell division (Dixit and Cyr 2002). The breakdown of the NE followed ruffling of the nuclear membranes in the area of the NE in clos-est proximity to the PPB. The spatio-temporal link between the loss of the PPB and breakdown of the NE in plant cells suggests that plant NE breakdown is initiated by membrane tearing and involves the attachment of MTs to the surface of the NE (Dixit and Cyr 2002).

A current topic of recent interest in the animal NE field is the process of NE re-formation at the end of mitosis. It has been known for some time that the NE assembles in a stepwise manner, with recruitment of INM proteins preceding that of pore proteins (Chaudhary and Courvalin 1993; Haraguchi et al. 2000). The finer mechanisms of this reassembly are now being elucidated. The INM protein SUN1 has been shown to be recruited at a very early stage to the condensed chromatin of the daughter nuclei (Chi et al. 2007). The protein was shown to associate with a histone acetyltransferase, in addition, when SUN1 expression was reduced, hyper acetylation of histones and delayed chromosome decondensation was observed. As such it suggests a possible functional role for SUN proteins in NE assembly.

Two recent publications using *Xenopus* egg extracts to study NE reassembly have dealt with the physical aspects of membrane recruitment and expansion to produce the new NEs. The recruitment of NE membranes around chromatin has been shown to arise from the intact tubular ER network in vitro (Anderson and Hetzer 2007). The ends of ER tubules were shown to bind directly to DNA and it is proposed that by virtue of the highly dynamic and mobile nature of the ER the membranes can quickly spread over the surface of the chromatin, with the membranes expanding and forming flattened sheets (Anderson and Hetzer 2007). In another study, Baur et al. (2007) demonstrated that, as for other membrane binding and fusion events in the cell, NSF and SNARE proteins are necessary for NE formation and successful pore complex assembly. As with many aspects of the plant NE in cell division there has been very little work undertaken on the process of plant NE reassembly; once again lack of native plant marker proteins is a clear problem.

Examples of location of native plant NE and NE-associated proteins during mitosis are gradually increasing. As previously mentioned immunolabelling of a tomato Ca-ATPase gave a strong NE signal, when studied in dividing cells the protein was found to localise in discrete domains within the mitotic membranes (Downie et al. 1998). This could suggest a possible role for the LCA-1 Ca-ATPase in regulating the calcium signalling pool in the mitotic apparatus (MA). Such targeting to specific regions of the MA raises interesting questions regarding the organisation of discrete domains within mitotic membranes.

Plant RanGAPs, the proteins that activate RanGTPase allowing the conversion of RanGTP to RanGDP, thus aiding the maintenance of the Ran gradient and directionality of nuclear import/export, are located at the outer side of the NE during interphase (Rose and Meier 2001; Pay et al. 2002). During mitosis these proteins are found associated with MTs within the mitotic apparatus (Pay et al. 2002), and are also associated with the phragmoplast at the end of division (Jeong et al. 2005). A specific N-terminal WPP domain in the plant RanGAP mediates the novel location of this protein during mitosis, which differs from that observed for animal cells (Jeong et al. 2005). An antibody raised against the *Arabidopsis* Ran2 protein also differed from the location seen for mammalian cells at interphase, the plant protein being found in the perinuclear and NE region but not in the nucleoplasm whereas the animal Ran proteins are usually present in the nucleus (Ma et al. 2007). These subtle differences in protein location between plant and animal Ran and Ran-associated proteins and kingdom-specific targeting domains highlights the fact that the plant NE clearly differs from that of its animal counterpart and as such provides a tantalising indication of the novel aspects of the plant nucleus that may be unveiled in the future.

10   Summary and Future Directions

Whilst the importance of the nucleus is well recognised in plant biology, research into its membranes and related protein constituents' remains in its infancy. Some information can be extrapolated from animal and yeast systems, but striking differences, such as the lack of plant homologues to lamins and to the vast majority of known animal nuclear membrane proteins point to the interesting possibility that the plant nuclear membranes contain novel proteins that contribute to an overall nuclear structure similar to that seen in other eukaryotic cells. Approaches using fluorescent proteins to localise novel candidates to the nuclear envelope and to describe their behaviour in processes involving it are proving fruitful in characterising this fascinating and important membrane structure. However, use of multiple approaches, including proteomics, mutagenesis and reverse genetics are needed to further the field. Given the role of the nuclear envelope and its interactions in processes as significant as cell division and in control of gene expression, the field must be considered an important one for future development.

Acknowledgements   S.I. is supported under the European Framework VI PharmaPlanta project and K.G. by an Oxford Brookes University research studentship.

References

Al-Mohanna FA, Caddy KWT, Bolsover SR (1994) The nucleus is insulated from large cytosolic calcium-ion changes. Nature 367:745–750

Anderson DJ, Hetzer MW (2007) Nuclear envelope formation by chromatin-mediated reorganization of the endoplasmic reticulum. Nat Cell Biol 9:1160–1166

Antonin W, Franz C, Haselmann U, Antony C, Mattaj IW (2005) The integral membrane nucleoporin pom121 functionally links nuclear pore complex assembly and nuclear envelope formation. Mol Cell 17:83–92

Baur T, Ramadan K, Schlundt A, Kartenbeck J, Meyer HH (2007) NSF- and SNARE-mediated membrane fusion is required for nuclear envelope formation and completion of nuclear pore complex assembly in *Xenopus laevis* egg extracts. J Cell Sci 120:2895–2903

Beaudouin J, Gerlich D, Daigle N, Eils R, Ellenberg J (2002) Nuclear envelope breakdown proceeds by microtubule-induced tearing of the lamina. Cell 108:83–96

Bengtsson L (2007) What MAN1 does to the Smads TGFb/BMP signaling and the nuclear envelope. FEBS J 274:1374–1382

Bootman MD, Thomas D, Tovey SC, Berridge MJ, Lipp P (2000) Nuclear calcium signalling. Cell Mol Life Sci 57:371–378

Bouche N, Yellin A, Snedden WA, Fromm H (2005) Plant-specific calmodulin-binding proteins. Ann Rev Plant Biol 56:435–466

Brandizzi F, Irons SL, Evans DE (2004) The plant nuclear envelope: new prospects for a poorly understood structure. New Phytol 163:227–246

Briere C, Xiong TC, Mazars C, Ranjeva R (2006) Autonomous regulation of free Ca^{2+} concentrations in isolated plant cell nuclei: a mathematical analysis. Cell Calcium 39:293–303

Broers JLV, Machiels BM, van Eys G, Kuijpers HJH, Manders EMM, van Driel R, Ramaekers FCS (1999) Dynamics of the nuclear lamina as monitored by GFP-tagged A-type lamins. J Cell Sci 112:3463–3475

Bunney TD, Shaw PJ, Watkins PAC, Taylor JP, Beven AF, Wells B, Calder GM, Drobak BL (2000) ATP-dependent regulation of nuclear Ca2+ levels in plant cells. FEBS Lett 476:145–149

Chaudhary N, Courvalin JC (1993) Stepwise reassembly of the nuclear-envelope at the end of mitosis. J Cell Biol 122:295–306

Chi Y-H, Haller K, Peloponese J-M Jr, Jeang K-T (2007) Histone acetyltransferase hALP and nuclear membrane protein hsSUN1 function in de-condensation of mitotic chromosomes. J Biol Chem 282:27447–27458. doi:10.1074/jbc.M703098200

Chial HJ, Rout MP, Giddings TH, Winey M (1998) Saccharomyces cerevisiae Ndc1p is a shared component of nuclear pore complexes and spindle pole bodies. J Cell Biol 143:1789–1800

Chikashige Y, Tsutsumi C, Yamane M, Kamasa K, Haraguchi T, Hiraoka Y (2006) Meiotic proteins Bqt1 and Bqt2 tether telomeres to form the bouquet arrangement of chromosomes. Cell 125:59–69

Chu A, Rassadi R, Stochaj U (1998) Velcro in the nuclear envelope: LBR and LAPs. FEBS Lett 441:165–169

Cohen M, Wilson KL, Gruenbaum Y (2001) Membrane proteins of the nuclear pore complex: Gp210 is conserved in *Drosophila*, *C. elegans* and Arabidopsis. In: Boulikas T (ed) Textbook of gene therapy and molecular biology: "from basic mechanism to clinical applications", vol. 6. Gene Therapy, Palo Alto, pp 47–55

Collings DA, Carter CN, Rink JC, Scott AC, Wyatt SE, Allen NS (2000) Plant nuclei can contain extensive grooves and invaginations. Plant Cell 12:2425–2440

Courvalin JC, Lassoued K, Bartnik E, Blobel G, Wozniak RW (1990) The 210-kD nuclear envelope polypeptide recognized by human autoantibodies in primary biliary cirrhosis is the major glycoprotein of the nuclear pore. J Clin Invest 86:279–285

Craig S, Staehelin LA (1988) High-pressure freezing of intact plant-tissues – evaluation and characterization of novel features of the endoplasmic-reticulum and associated membrane systems. Eur J Cell Biol 46:80–93

Crisp M, Liu Q, Roux K, Rattner JB, Shanahan C, Burke B, Stahl PD, Hodzic D (2006) Coupling of the nucleus and cytoplasm: role of the LINC complex. J Cell Biol 172:41–53

Demidov D, Van Damme D, Geelen D, Blattner FR, Houben A (2005) Identification and dynamics of two classes of aurora-like kinases in Arabidopsis and other plants. Plant Cell 17:836–848

Dixit R, Cyr RJ (2002) Spatio-temporal relationship between nuclear-envelope breakdown and preprophase band disappearance in cultured tobacco cells. Protoplasma 219:116–121

Downie L, Priddle J, Hawes C, Evans DE (1998) A calcium pump at the higher plant nuclear envelope? FEBS Lett 429:44–48

Echevarria W, Leite MF, Guerra MT, Zipfel WR, Nathanson MH (2003) Regulation of calcium signals in the nucleus by a nucleoplasmic reticulum. Nat Cell Biol 5:440–446

Ellenberg J, Siggia ED, Moreira JE, Smith CL, Presley JF, Worman HJ, Lippincott-Schwartz J (1997) Nuclear membrane dynamics and reassembly in living cells: targeting of an inner nuclear membrane protein in interphase and mitosis. J Cell Biol 138:1193–1206

Fava F, Raynaud-Messina B, Leung-Tack J, Mazzolini L, Li M, Guillemot JC, Cachot D, Tollon Y, Ferrara P, Wright M (1999) Human 76p: a new member of the gamma-tubulin-associated protein family. J Cell Biol 147:857–868

Foisner R (2001) Inner nuclear membrane proteins and the nuclear lamina. J Cell Sci 114:3791–3792

Fricker M, Hollinshead M, White N, Vaux D (1997) Interphase nuclei of many mammalian cell types contain deep, dynamic, tubular membrane-bound invaginations of the nuclear envelope. J Cell Biol 136:531–544

Gerace L, Burke B (1988) Functional organization of the nuclear envelope. Ann Rev Cell Biol 4:335–374

Graumann K, Irons SL, Runions J, Evans DE (2007) Retention and mobility of the mammalian lamin B receptor in the plant nuclear envelope. Biol Cell 99:553–562

Greber UF, Senior A, Gerace L (1990) A major glycoprotein of the nuclear pore complex is a membrane-spanning polypeptide with a large lumenal domain and a small cytoplasmic tail. EMBO J 9:1495–1502

Gruenbaum Y, Margalit A, Goldman RD, Shumaker DK, Wilson KL (2005) The nuclear lamina comes of age. Nat Rev 6:21–31

Grygorczyk C, Grygorczyk R (1998) A Ca2+- and voltage-dependent cation channel in the nuclear envelope of red beet. Biochim Biophys Acta-Biomembranes 1375:117–130

Haraguchi T, Koujin T, Segura M, Wilson KL, Hiraoka Y (2000) Dynamic behavior of emerin and BAF at early stages of nuclear assembly in living HeLa cells. Mol Biol Cell 11:21A

Hofemeister H, O'Hare P (2005) Analysis of the location and topology of nurim, a polytopic protein tightly associated with the inner nuclear membrane. J Biol Chem 280:2512–2521

Irons SL, Evans DE, Brandizzi F (2003) The first 238 amino acids of the human lamin B receptor are targeted to the nuclear envelope in plants. J Exp Bot 54:943–950

Jeong SY, Rose A, Joseph J, Dass M, Meier I (2005) Plant-specific mitotic targeting of RanGAP requires a functional WPP domain. Plant J 42:270–282

Kalo P, Gleason C, Edwards A, Marsh J, Mitra RM, Hirsch S, Jakab J, Sims S, Long SR, Rogers J, Kiss GB, Downie JA, Oldroyd GED (2005) Nodulation signaling in legumes requires NSP2, a member of the GRAS family of transcriptional regulators. Science 308:1786–1789

Kawabe A, Matsunaga S, Nakagawa K, Kurihara D, Yoneda A, Hasezawa S, Uchiyama S, Fukui K (2005) Characterization of plant Aurora kinases during mitosis. Plant Mol Biol 58:1–13

Kemp CA, Song MH, Addepalli MK, Hunter G, O'Connell K (2007) Suppressors of zyg-1 define regulators of Centrosome Duplication and Nuclear Association in Caenorhabditis elegans. Genetics 176:95–113

King MC, Lusk CP, Blobel G (2006) Karyopherin-mediated import of integral inner nuclear membrane proteins. Nature 442:1003–1007

Lagace TA, Ridgway ND (2005) Induction of apoptosis by lipophilic activators of CTP: phospho-choline cytidylyltransferase alpha (CCT alpha). Biochem J 392:449–456

Ledeen RW, Wu G (2007) Sodium-calcium exchangers in the nucleus: an unexpected locus and unusual regulatory mechanism. Ann NY Acad Sci 1099:494–506

Lui PPY, Lee CY, Tsang D, Kong SK (1998) Ca^{2+} is released from the nuclear tubular structure into nucleoplasm in C6 glioma cells after stimulation with phorbol ester. FEBS Lett 432:82–87

Lusk CP, Blobel G, King MC (2007) Highway to the inner nuclear membrane: rules for the road. Nat Rev Mol Cell Biol 8:414–420

Ma Y, Cai S, Lv QL, Jiang Q, Zhang Q, Sodmergen, Zhai ZH, Zhang CM (2007) Lamin B receptor plays a role in stimulating nuclear envelope production and targeting membrane vesicles to chromatin during nuclear envelope assembly through direct interaction with importin beta. J Cell Sci 120:520–530

Mans BJ, Anantharaman V, Aravind L, Koonin EV (2004) Comparitive genomics, evolution and origins of the nuclear envelope and nuclear pore complex. Cell Cycle 3:1612–1637

Mansfeld J, Güttinger S, Hawryluk-Gara LA, Panté N, Mall M, Galy V, Haselmann U, Mühlhäusser P, Wozniak RW, Mattaj IW, Kutay U, Antonin W (2006) The conserved transmembrane nucleoporin NDC1 is required for nuclear pore complex assembly in vertebrate cells. Mol Cell 22:93–103

Mansharamani M, Hewetson A, Chilton BS (2001) Cloning and characterisation of an atypical type IV P-type ATPase that binds to the RING motif of RUSH transcription factors. J Biol Chem 276:3641–3649

Marius P, Guerra MT, Nathanson MH, Ehrlich BE, Leite MF (2006) Calcium release from ryanodine receptors in the nucleoplasmic reticulum. Cell Calcium 39:65–73

Mattaj IW (2004) Sorting out the nuclear envelope from the endoplasmic reticulum. Nat Rev Mol Cell Biol 5:65–69

Meier I (2001) The plant nuclear envelope. Cell Mol Life Sci 58:1774–1780

Merkle T (2008) Nuclear export of protein and RNA Plant Cell Monogr doi: 10.1007/7089_2008_25

Miao M, Ryan KJ, Wente SR (2006) The integral membrane protein Pom34p functionally links nucleoporin subcomplexes. Genetics 172:1441–1457

Morena Diaz de la Espina S (2008) The nucleoskeleton. Plant Cell Monogr doi:10.1007/7089_2008_26

Mosley-Bishop KL, Li Q, Patterson K, Fischer JA (1999) Molecular analysis of the klarsicht gene and its role in nuclear migration within differentiating cells of the Drosophila eye. Curr Biol 9:1211–1220

Murphy SM, Preble AM, Patel UK, O'Connell KL, Dias DP, Moritz M, Agard D, Stults JT, Stearns T (2001) GCP5 and GCP6: two new members of the human gamma-tubulin complex. Mol Biol Cell 12:3340–3352

Nikolakaki E, Meier J, Simos G, Georgatos SD, Giannakouros T (1997) Mitotic phosphorylation of the lamin B receptor by a serine/arginine kinase and p34(cdc2). J Biol Chem 272:6208–6213

Ohba T, Schirmer EC, Nishimoto T, Gerace L (2004) Energy- and temperature-dependent transport of integral proteins to the inner nuclear membrane via the nuclear pore. J Cell Biol 167:1051–1062

Padmakumar VC, Abraham S, Braune S, Noegel AA, Tunggal B, Karakesisoglou I, Korenbaum E (2004) Enaptin, a giant actin-binding protein, is an element of the nuclear membrane and the actin cytoskeleton. Exp Cell Res 295:330–339

Pauly N, Knight MR, Thuleau P, van der Luit AH, Moreau M, Trewavas AJ, Ranjeva R, Mazars C (2000) Cell signalling – control of free calcium in plant cell nuclei. Nature 405:754–755

Pay A, Resch K, Frohnmeyer H, Fejes E, Nagy F, Nick P (2002) Plant RanGAPs are localized at the nuclear envelope in interphase and associated with microtubules in mitotic cells. Plant J 30:699–709

Peiter E, Sun J, Heckmann AB, Venkateshwaran M, Riley BK, Otegui MS, Edwards A, Freshour G, Hahn MG, Cook DR, Sanders D, Oldroyd GED, Downie JA, Ané J-M (2007) The Medicago truncatula DMI1 Protein Modulates Cytosolic Calcium Signaling. Plant Physiol Preview doi:10.1104/pp.107.097261

Prunuske A, Ullman K (2006) The nuclear envelope: form and reformation. Curr Opin Cell Biol 18:108–116

Riely BK, Lougnon G, Ane JM, Cook DR (2006) The symbiotic ion channel homolog DMI1 is localized in the nuclear membrane of Medicago truncatula roots. Plant J 49:208–216

Rose A (2008) Nuclear pores in plant cells: structure, composition, and functions plant. Cell Monogr doi:10.1007/7089_2008_27

Rose A, Meier, I (2001) A domain unique to plant RanGAP is responsible for its targeting to the plant nuclear rim. Proc Natl Acad Sci USA 98:15377–15382

Rose A, Patel S, Meier I (2004) Plant nuclear envelope proteins. Symp Soc Exp Biol 69–88

Saksena S, Shao Y, Braunagel SC, Summers MD, Johnson AE (2004) Cotranslational integration and initial sorting at the endoplasmic reticulum translocon of proteins destined for the inner nuclear membrane. Proc Natl Acad Sci USA 101:12537–12542

Saksena S, Summers MD, Burks JK, Johnson AE, Braunagel SC (2006) Importin-alpha-16 is a translocon-associated protein involved in sorting membrane proteins to the nuclear envelope. Nat Struct Mol Biol 13:500–508

Salina D, Bodoor K, Eckley DM, Schroer TA, Rattner JB, Burke B (2002) Cytoplasmic dynein as a facilitator of nuclear envelope breakdown. Cell 108:97–107

Santella L, Kyozuka K (1997) Effects of 1-methyladenine on nuclear Ca^{2+} transients and meiosis resumption in starfish oocytes are mimicked by the nuclear injection of inositol 1,4,5-trisphosphate and cADP-ribose. Cell Calcium 22:11–20

Schirmer EC, Gerace L (2005) The nuclear membrane proteome: extending the envelope. Trends Biochem Sci 30:551–558

Schmitt J, Benavente R, Hodzic D, Hoog C, Stewarts CL, Alsheimer M (2007) Transmembrane protein Sun2 is involved in tethering mammalian meiotic telomeres to the nuclear envelope. Proc Natl Acad Sci USA 104:7426–7431

Seltzer V, Janski N, Canaday J, Herzog E, Erhardt, M, Evrard JL, Schmit AC (2007) Arabidopsis GCP2 and GCP3 are part of a soluble gamma-tubulin complex and have nuclear envelope targeting domains. Plant Journal 52:322–331

Shimamura M, Brown RC, Lemmon BE, Akashi T, Mizuno K, Nishihara N, Tomizawa KI, Yoshimoto K, Deguchi H, Hosoya H, Horio T, Mineyuki Y (2004) gamma-Tubulin in basal land plants: Characterization, localization, and implication in the evolution of acentriolar microtubule organizing centers. Plant Cell 16:45–59

Staehelin LA (1997) The plant ER: a dynamic organelle composed of a large number of discrete functional domains. Plant J 11:1151–1165

Starr DA, Fischer JA (2005) KASH 'n Karry: the KASH domain family of cargo-specific cytoskeletal adaptor proteins. Bioessays 27:1136–1146

Starr DA, Han M (2003) ANChors away: an actin based mechanism of nuclear positioning. J Cell Sci 116:211–216

Stavru F, Hülsmann BB, Spang A, Hartmann E, Cordes VC, Görlich D (2006) NDC1: a crucial membrane-integral nucleoporin of metazoan nuclear pore complexes. J Cell Biol 173:509–519

Stoppin V, Lambert AM, Vantard M (1996) Plant microtubule-associated proteins (MAPs) affect microtubule nucleation and growth at plant nuclei and mammalian centrosomes. Eur J Cell Biol 96:11–23

Tomita K, Cooper JP (2006) The meiotic chromosomal bouquet: SUN collects flowers. Cell 125:19–21

Tzur YB, Margalit A, Melamed-Book N, Gruenbaum Y (2006a) Matefin/SUN-1 is a nuclear envelope receptor for CED-4 during Caenorhabditis elegans apoptosis. Proc Natl Acad Sci USA 103:13397–13402

Tzur YB, Wilson KL, Gruenbaum Y (2006b) SUN-domain proteins: 'Velcro' that links the nucleoskeleton to the cytoskeleton. Nat Rev Mol Cell Biol 7:782–788

Van Damme D, Bouget FY, Van Poucke K, Inze D, Geelen D (2004) Molecular dissection of plant cytokinesis and phragmoplast structure: a survey of GFP-tagged proteins. Plant J 40:386–398

Wilhelmsen K, Litjens SH, Kuikman M, Tshimbalanga N, Janssen H, van den Bout I, Raymond K, Sonnenberg A (2005) Nesprin-3, a novel outer nuclear membrane protein, associates with the cytoskeletal linker protein plectin. J Cell Biol 171:799–810

Wilhelmsen K, Ketema M, Truong H, Sonnenberg A (2006) KASH-domain proteins in nuclear migration, anchorage and other processes. J Cell Sci 119:5021–5029

Wilkie GS, Schirmer EC (2006) Guilt by association: the nuclear envelope proteome and disease. Mol Cell Proteomics 5:1865–1875

Winey M, Hoyt MA, Chan C, Goetsch L, Botstein D, Byers B (1993) NDC1: a nuclear periphery component required for yeast spindle pole body duplication. J Cell Biol 122:743–751

Worman HJ, Gundersen GG (2006) Here come the SUNs: a nucleocytoskeletal missing link. Trends Cell Biol 16:67–69

Wozniak RW, Blobel G (1992) The single transmembrane segment of gp210 is sufficient for sorting to the pore membrane domain of the nuclear envelope. J Cell Biol 119:1441–1449

Xiong TC, Jauneau A, Ranjeva R, Mazars C (2004) Isolated plant nuclei as mechanical and thermal sensors involved in calcium signalling. Plant J 40:12–21

Xu XM, Meulia T, Meier I (2007) Anchorage of Plant RanGAP to the nuclear envelope involves novel nuclear-pore-associated proteins. Curr Biol 17:1157–1163

Yang L, Guan TL, Gerace L (1997) Lamin-binding fragment of LAP2 inhibits increase in nuclear volume during the cell cycle and progression into S phase. J Cell Biol 139:1077–1087

Zargari A, Boban M, Heessen S, Andréasson C, Thyberg J, Ljungdahl PO (2007) Inner nuclear membrane proteins Asi1, Asi2, and Asi3 function in concert to maintain the latent properties of transcription factors Stp1 and Stp2. J Biol Chem 282:594–605

Zhang Q, Ragnauth C, Greener MJ, Shanahan C, Roberts RG (2002) The nesprins are giant actin-binding proteins, orthologous to *Drosophila melanogaster* muscle protein MSP-300. Genomics 80:473–481

Nuclear Pores in Plant Cells: Structure, Composition, and Functions

Annkatrin Rose

Abstract The nuclear pores form the gateways connecting the nucleoplasm of eukaryotic cells with the cytoplasm. They are essentially fusions of the inner and outer nuclear membranes forming a connecting pore membrane and a "hole" in the nuclear envelope. They are organized and anchored by a multi-protein complex termed the nuclear pore complex (NPC), which facilitates and regulates the transport of molecules across the barrier provided by the nuclear envelope. While proteomics analysis has unraveled the molecular composition of the vertebrate and yeast NPC, our knowledge of the plant nuclear pore is still far from comprehensive. However, several components of the plant NPC and nucleocytoplasmic transport machinery have emerged in recent mutant screens and were found to affect diverse processes ranging from plant–microbe interactions and hormone and stress responses to development and the regulation of flowering time. Taken together, these studies illustrate the importance of the NPC and nucleocytoplasmic transport in the regulation of plant growth and responses to the environment.

1 Structure of the Nuclear Pore

Nuclear pores were first discovered nearly six decades ago via electron microscopy of the isolated nuclear envelopes of amphibian oocytes (Callan and Tomlin 1950). Subsequent studies in the 1960s and early 1970s demonstrated that the microscopic structure of plant nuclear pores resembles that of animal pores (Roberts and Northcote 1970). In all three kingdoms of eukaryotes, the microscopic structure of the nuclear pores shows an eightfold symmetry, and its overall architecture appears to be evolutionarily conserved although the overall dimension of the NPC may vary between species (Evans et al. 2004; Heese-Peck and Raikhel 1998; Lim et al. 2008). The pore structure has been most extensively studied in the *Xenopus laevis*

A. Rose
Department of Biology, Appalachian State University, 572 Rivers Street, Boone, NC 28608, USA
e-mail: rosea@appstate.edu

Plant Cell Monogr, doi:10.1007/7089_2008_27

oocyte system. In electron micrographs, *Xenopus* nuclear pores appear with an overall cylindrical shape of 120-nm diameter and 70 nm in height spanning the nuclear envelope. They exhibit an eightfold rotational symmetry viewed from the top and a tripartite asymmetrical structure in a side view (Heese-Peck and Raikhel 1998; Reichelt et al. 1990). NPCs from both mono- and dicotyledonous plants appear structured according to the same symmetrical arrangements (Heese-Peck and Raikhel 1998; Roberts and Northcote 1970).

The central part of the pore contains eight spokes known as the spoke complex connecting a cytoplasmic and a nucleoplasmic ring. Rod-like fibrils extend from the cytoplasmic ring into the cytoplasm, while a nucleoplasmic basket extends from the nuclear ring. These structures are thought to serve as docking sites for import and export substrates, respectively (Heese-Peck and Raikhel 1998; Panté and Aebi 1996). Filaments extending from the nuclear basket into the nucleoplasm link neighboring NPCs together (Ris 1997). The NPC core forms a central channel with a functional diameter of up to 26–39 nm and eight smaller channels with 8–9-nm diameter that allow the passive diffusion of small molecules (Feldherr and Akin 1990; Hinshaw et al. 1992; Panté and Kann 2002).

2 Molecular Composition of the Nuclear Pore

Nuclear pores connect the outer with the inner nuclear membrane, thus forming three distinct membrane domains within the NPC which can be distinguished functionally and biochemically (Drummond and Allen 2004). The outer nuclear membrane (ONM) is continuous with the endoplasmic reticulum. In metazoans, the ONM contains specific proteins, such as the spectrin-like nesprins that connect the nucleus with the actin cytoskeleton (Zhang et al. 2001), while the inner nuclear membrane (INM) is attached to the nuclear lamina, a meshwork of lamin intermediate filaments and interacting proteins underlying the nuclear envelope. This protein meshwork provides a means of association of the INM with the nuclear matrix and chromatin inside the nucleus (Goldberg et al. 1999). Nesprins and lamins appear to be unique to metazoans and no homologs have been identified in plants (Mans et al. 2004). The pore membrane (POM) constitutes the fusion membrane between the ONM and INM and contains pore-specific integral membrane proteins that facilitate assembly and anchoring of the NPC in the nuclear envelope (Gerace and Foisner 1994).

On a molecular level, the nuclear pore complex is a multi-protein complex of about 125 MDa in size in animals and about 50–80 MDa in yeast cells (Evans et al. 2004). Despite the size difference, the nuclear pores of yeast and mammals share many similar morphological features. Proteomic studies in animals and yeast have revealed about 30 proteins, termed nucleoporins (Nups) and typically referred to by their molecular weight, as building blocks of the nuclear pore complex (Cronshaw et al. 2002; Rout et al. 2000). The larger size of the vertebrate nuclear pore results from multiple copies of these proteins rather than increased complexity of the NPC proteome.

Many nucleoporins occur in copies of eight, reflecting the eightfold symmetry of the nuclear pore observed in electron micrographs. Some nucleoporins preferentially occur on either the cytoplasmic or nucleoplasmic side of the pore, while most are arranged symmetrically on both sides (Dreger and Otto 2004).

The composition of the plant nuclear pore complex is still largely unknown. For about half of the nucleoporins identified in the metazoan NPC, sequence similarity searches have failed to identify plant homologs. Only half of the nucleoporins identified so far in plants have been characterized experimentally while the remaining proteins are only inferred through sequence similarity with animal or yeast nucleoporins (Xu and Meier 2007). For a summary of the known *Arabidopsis* nucleoporins, see Table 1.

2.1 Nuclear Pore Subcomplexes and Plant Nucleoporins

The nuclear pore complexes are anchored in the pore membrane via integral pore membrane proteins (Poms). In vertebrates, the proteins Pom210 (also referred to as gp210 for glycoprotein 210), Pom121, and NDC1 are involved in the anchoring of the nuclear pore and reformation of nuclear pores after mitosis (Cronshaw et al. 2002; Mansfeld et al. 2006). Three anchoring integral membrane proteins have been identified in yeast, Pom34p, Pom152p, and Ndc1p (Rout et al. 2000). Pom152p is similar to Pom210/gp210 and in addition to anchoring the nuclear pore complex also appears to be an integral part of the nuclear envelope-embedded spindle pole body in yeast (Chial et al. 1998). A putative ortholog of Pom210/gp210 has been identified in plants and is predicted to contain a C-terminal transmembrane domain (Cohen et al. 2001). In addition to Pom210/gp210, conservation of NDC1 in plants can be inferred from sequence similarity (Xu and Meier 2007).

Seven additional proteins form the spoke ring and the central transport channel of the NPC core. Compared with the metazoan NPC, only one protein of the central channel, Nup62, and none in the spoke complex appear to have sequence homologs in plants (Xu and Meier 2007). This central core is flanked on both sides of the nuclear envelope by symmetrical cytoplasmic and nuclear rings, which share the same nucleoporin composition of roughly a dozen different proteins. The majority of proteins in these coaxial ring complexes appears conserved in plants, with the exception of Nup37, Nup214, and Seh1 (Xu and Meier 2007). In the metazoan NPC, this complex, also referred to as the Nup107-160 complex, is considered to form a central scaffold of the pore required for nuclear pore assembly (Harel et al. 2003b; Walther et al. 2003a). Interestingly, while depletions of nucleoporins in this complex severely reduce the number of pores per nucleus in animal cells and mutations are often lethal, several viable *Arabidopsis* and lotus mutants have recently been isolated that involve plant nucleoporins putatively located within these symmetrical ring complexes (Dong et al. 2006; Kanamori et al. 2006; Parry et al. 2006; Saito et al. 2007; Zhang and Li 2005). The comparatively mild effect of these mutations in plants is surprising and might indicate yet to be uncovered redundancy in

Table 1 Plant nucleoporins, associated mutant phenotypes and putative orthologs

Protein	*Arabidopsis* gene(s)[a]	Putative location[a]	Sequence motifs	Mutant phenotypes	Mammalian ortholog	Yeast ortholog[b]
WIP1, 2, 3	AT4G26450 AT5G56210 AT3G13360	ONM/POM	TMD, coiled-coil	Loss of RanGAP attachment to the NPC in triple mutants (Xu et al. 2007a)	None	None
gp210/ EMB3012	AT5G40480	POM	TMD	Embryo defective: arrest at pre-globular stage[c]	Gp210	None
NDC1	AT1G73240	POM	TMD		NDC1 (Stavru et al. 2006)	Ndc1p
ALADIN	AT3G56900	Cytoplasmic fibrils	WD repeat		ALADIN	None
Nup88/MOS7	unpublished	Cytoplasmic fibrils		Suppression of *snc1*, defects in basal and systemic acquired resistance (Wiermer et al. 2007)	Nup88	Nup82p
Nup98	AT1G10390 AT1G59660	Coaxial rings	FG repeat		Nup98	Nup145N, Nup116p, Nup100p
Nup160/SAR1	AT1G33410	Coaxial rings, Nup107-160		Suppression of *axr1*, early flowering, retarded growth (Cernac et al. 1997; Dong et al. 2006; Parry et al. 2006)	Nup160	Nup120p
Nup96/MOS3/ SAR3/PRE	AT1G80680	Coaxial rings, Nup107-160		Suppression of *snc1* and *axr1*, early flowering (Parry et al. 2006; Zhang and Li 2005)	Nup96	Nup145C
Nup85	AT4G32910	Coaxial rings, Nup107-160		Allele-specific defects in symbiotic interactions, reduced number of seeds (Saito et al. 2006)[d]	Nup75	Nup85p
Nup107	AT3G14120	Coaxial rings, Nup107-160	Leu zipper		Nup107	Nup84p

Nup133	AT2G05120	Coaxial rings, Nup107-160		Temperature-sensitive nodulation deficiency, absence of mycorrhiza, reduced number of seeds (Kanamori et al. 2006)[d]	Nup133	Nup133p
Nup43	AT4G30840	Coaxial rings, Nup107-160	WD repeat		Nup43	None
Sec13	AT2G30050 AT3G01340	Coaxial rings			Sec13R	Sec13p
Nup62/ EMB2766	AT2G45000 EMB2766	Central channel	FG repeat, coiled-coil	Embryo defective[c]	Nup62	Nsp1p
Nup155	AT1G14850	Nuclear basket			Nup155	Nup157p, Nup170p
Tpr/NUA	AT1G79280	Nuclear basket	Coiled-coil	Early flowering, stunted growth, changes in phyllotaxy, reduced fertility, suppression of axr1 (Jacob et al. 2007; Xu et al. 2007b)	Tpr	Mlp1p, Mlp2p

[a]Putative locations within the NPC and components of the Nup107-160 complex are inferred from known locations and interactions of vertebrate homologs; nuclear envelope localization has been confirmed experimentally for the following proteins: WIP family (Xu et al. 2007a), Nup160 (Dong et al. 2006), MOS3/SAR3 (Parry et al. 2006; Zhang and Li 2005), Nup133 (Kanamori et al. 2006), and NUA (Xu et al. 2007b)

[b]Yeast orthologs of mammalian Nups according to Cronshaw et al. 2002, unless cited otherwise

[c]Phenotype curated by ABRC (www.arabidopsis.org(based on results from the SeedGenes project (Tzafrir et al. 2003)

[d]Phenotype observed in Lotus japonicus; all other phenotypes observed in Arabidopsis

ONM, outer nuclear membrane, POM, pore membrane, TMD, transmembrane domain the plant NPC or that the involved mutant alleles are not null alleles and retain partial functions of the affected nucleoporins (Xu and Meier 2007).

On the nucleoplasmic side of the pore, a nuclear pore basket extends from the nuclear coaxial ring. In metazoan cells, the filaments of the nuclear basket are formed by the long coiled-coil protein Tpr (Cordes et al. 1997; Paddy 1998; Zimowska et al. 1997), which is connected to the nuclear coaxial ring via the nucleoporin Nup153 (Krull et al. 2004). Nup153 also serves as an anchor for the SUMO protease SENP2 (Hang and Dasso 2002). Other proteins of the nuclear basket in metazoan cells include Nup155 and Nup50. Of these proteins, only Tpr and Nup155 appear conserved in plants (Xu and Meier 2007). Loss-of-function alleles of Nup 154, the Drosophila homolog of Nup155, are lethal while milder alleles cause sterility (Kiger et al. 1999). Mutations in *Arabidopsis* Tpr (or NUA for NUCLEAR PORE ANCHOR) have pleiotropic phenotypes similar to those caused by mutants affecting the coaxial ring complexes. Because no homolog of Nup153 exists in plants, it is unclear how plant Tpr is anchored at the pore. In *Arabidopsis*, the SUMO protease ESD4 interacts directly with NUA/Tpr (Xu et al. 2007b), thus linking the SUMO pathway to the nuclear basket of the plant NPC but in a manner different from mammalian cells and more similar to the interaction between the SUMO protease Ulp1p and the Tpr homologs Mlp1p and Mlp2p in yeast (Zhao et al. 2004).

On the other side of the pore, cytoplasmic fibrils extend from the cytoplasmic coaxial ring. The protein composition of these is even less conserved between metazoans and plants than that of the nuclear basket. Only Nup88 was found to have a homolog in plants and was isolated in a screen for mutants defective in plant pathogen response (Wiermer et al. 2007). The cytoplasmic fibrils of the mammalian NPC contain RanBP2 (Nup358), which serves as the anchor for RanGAP attachment (Mahajan et al. 1997). No RanBP2 sequence homolog exists in plants or yeast, however plant nuclear pores contain another group of nucleoporins termed WPP-domain Interacting Proteins (WIPs) which anchor RanGAP to the plant nuclear pore (Xu et al. 2007a). As with the SUMO pathway, a connection between the Ran cycle and the nuclear pore exists in plants as well as animals, however utilizing different mechanisms of attachment suggestive of convergent evolution.

The exact molecular composition of the plant NPC is not yet known, and it is therefore possible that additional plant Nups remain to be discovered. It will be interesting to see whether the plant NPC contains plant-specific Nups to "fill the gaps" in the complexes lacking homologs to mammalian or yeast proteins. The only plant-specific nucleoporins identified so far are the putative nucleoporins of the WIP family involved in RanGAP attachment (Xu et al. 2007a).

2.2   *Functional Motifs and Modifications*

Nucleoporins can be classified into families sharing specific sequence or structure motifs, which often suggest a common function. A subset of nucleoporins contains phenylalanine- and glycine-rich sequences in the form of FG repeats (Dreger and Otto 2004). It has been estimated that there are approximately 10,000 of these FG repeats or similar hydrophobic motifs in each NPC (Drummond and Allen 2004). These FG nucleoporins are thought to act as transient, low-affinity binding sites for the attachment of transport complexes trafficking through the nuclear pore (Mattaj and Englmeier 1998) and have been shown to be natively unfolded to provide flexibility to the transport channel of the pore (Denning et al. 2003; Lim et al. 2008). The direct interaction of transport factors with these phenylalanine-rich domains is essential for the translocation through the pore (Bayliss et al. 1999; Fribourg et al. 2001; Strasser et al. 2000). Putative orthologs of the FG-repeat-containing nucleoporins Nup98 and Nup62 can also be identified via sequence similarity in plant genomes (Rose et al. 2004; Xu and Meier 2007).

Another motif recurrent in nucleoporins is the WD repeat, which consists of an about 40–60 amino acid long repeat ending in WD (tryptophan-aspartate) and is found in a wide variety of proteins in multiple cellular contexts. On the basis of crystallography, the WD repeat is likely to adopt a circularized beta-propeller structure (Drummond and Allen 2004; Li and Roberts 2001) and act as a protein–protein interaction platform in the assembly of protein complexes (van Nocker and Ludwig 2003). Several WD repeat proteins have been identified as components of the vertebrate NPC and a homolog of the mammalian WD nucleoporin Nup43 can also be identified in the *Arabidopsis* genome (Xu and Meier 2007).

Nucleoporins in animals and plants may be glycosylated. In both plants and animals, N-acetylglucosamine (GlcNAc) residues were identified as nucleoporin modifications and have been used to isolate putative nucleoporins via specific interaction with wheat germ agglutinin (Heese-Peck and Raikhel 1998; Miller et al. 2000). It has been suggested that the possible function of this modification may involve competition with phosphorylation as hyperglycosylation caused by blocking the enzymatic removal of GlcNAc residues leads to hypophosphorylated proteins (Haltiwanger et al. 1998). In the context of the mammalian nuclear pore, several glycosylated nucleoporins such as Nup153, Nup214, Nup358 and gp210 become hyperphosphorylated during mitosis, thus disrupting protein–protein interactions in the disassembly of NPCs (Favreau et al. 1996; Macaulay et al. 1995). In contrast to animal nucleoporins that contain just one GlcNAc residue, the tobacco nucleoporin gp40 contains chains of five or more residues (Heese-Peck et al. 1995). The interaction of WGA with nucleoporins inhibits nuclear protein import in vertebrates, but only affects the import of large complexes in plants (Hicks et al. 1996). The longer chain length of the GlcNAc polysaccharide providing more space for smaller proteins to pass between the NPC and WGA in plant cells might offer a possible explanation for this phenomenon (Heese-Peck et al. 1995). However, rice importin α1a can release the WGA inhibition of nuclear protein import in permeabilized HeLa cells, suggesting a direct link to plant importin α rather than the NPC (Yamamoto and Deng 1999). The functional significance of the different glycosylation chain lengths in animal and plant cells is not known.

In mammalian and yeast cells, the nucleoporins Nup98/N-Nup145p and Nup96/C-Nup145p, respectively, are generated through the autocatalytic cleavage of a larger poly-protein precursor (Rosenblum and Blobel 1999; Teixeira et al. 1997). This autoproteolysis is essential for the localization of both proteins (Fontoura et al. 1999). In contrast to this processing step conserved in yeast and animals, plants appear to encode homologs of these two nucleoporins as separate polypeptides. SAR3/MOS3, the *Arabidopsis* homolog of Nup96, encompasses a region of homology to the entire Nup96 protein as well as the C-terminal 196 amino acids of Nup98 at the N-terminus of SAR3/MOS3 (Parry et al. 2006). In addition, the *Arabidopsis* genome also encodes two proteins related to Nup98 (At1g10390 and At1g59660). Interestingly, all three polypeptides contain the proteolytic motif present at the C-terminus of mammalian Nup98, but are encoded by separate genes in *Arabidopsis* (Parry et al. 2006). It is unclear whether any of the *Arabidopsis* nucleoporins undergo autoproteolytic cleavage as a form of protein maturation.

3 Functions of the Plant Nuclear Pore Complex

The NPC primarily functions as the site of nucleocytoplasmic transport. Recent studies have provided an increasing body of evidence for diverse functions of nuclear pores in plant cells. Genetic screens have identified nucleoporins and components of the transport machinery to be involved in diverse processes including plant–microbe interactions, auxin response, tolerance to cold stress, and the regulation of flowering time (Dong et al. 2006; Jacob et al. 2007; Kanamori et al. 2006; Parry et al. 2006; Saito et al. 2007; Xu et al. 2007b; Zhang and Li 2005).

3.1 Nucleocytoplasmic Transport

The main function of the nuclear pores is to allow for the diffusion of small molecules and the active transport of macromolecules across the nuclear envelope. In human HeLa cells, the rate of transport across the nuclear pore has been measured at ten import and ten export events per second per pore, which corresponds to about 600,000 ribosomes being transported per nucleus per minute (Görlich and Kutay 1999; Macara 2001; Mattaj and Englmeier 1998). Molecules smaller than 30 kDa can readily diffuse through the nuclear pores (Gasiorowski and Dean 2003; Stoffler et al. 1999), and passive transport across the pore is possible for molecules up to 60 kDa (Gerace and Burke 1988). The transport of larger molecules requires active transport across the pore. This transport system through the pore is directed by the Ran cycle and transport receptors called karyopherins in combination with nuclear localization or export signals on the molecules being transported (Pemberton and Paschal 2005). The components controlling Ran-mediated active nucleocytoplasmic transport appear conserved in all eukaryotic kingdoms and some have been shown to function in heterologous systems. Three major types of nuclear localization signals (NLS) identified in animals also function in plants and are recognized by *Arabidopsis* importin α (Smith and Raikhel 1999; Smith et al. 1997). Components of the Ran cycle and karyopherin homologs have been identified in plants and shown to complement the corresponding yeast mutants (Ach and Gruissem 1994;

Ballas and Citovsky 1997; Hunter et al. 2003; Merkle et al. 1994; Pay et al. 2002). This suggests a high degree of evolutionary conservation for the basic machinery responsible for nucleocytoplasmic transport. Translocation of the transport complex is thought to involve interaction of the import complex with FG-repeat containing nucleoporins and an affinity gradient for importin β across the NPC (Ben-Efraim and Gerace 2001; Lim et al. 2008).

3.1.1   Nuclear Import in Plants

Nuclear import in plants shows unique features in contrast to other eukaryotes. In animals, importin α serves as the receptor for the NLS on the cargo protein while importin β interacts with the NPC. Thus, the formation of a ternary complex of cargo-importin α-importin β in the cytoplasm precedes and is required for facilitating transport across the nuclear envelope (Merkle 2001; Pemberton and Paschal 2005; Stewart 2007). In contrast to animal systems where depletion of the cytoplasm from perforated cells abolishes nuclear import due to depletion of import factors, protoplasts from higher plants were found to retain importin α and their ability for protein import when permeabilized (Hicks et al. 1996; Merkle et al. 1996). Further studies uncovered that plant importin α is strongly associated with the nuclear envelope (Hicks et al. 1996; Smith et al. 1997) and is sufficient to mediate protein import into rat nuclei independently of interaction with importin β (Hübner et al. 1999).

3.1.2   The Ran Cycle

Nucleocytoplasmic transport is facilitated by the Ran cycle, which provides the molecular signals to distinguish between cytoplasm and nucleoplasm. This distinction takes the form of a sharp concentration gradient of the GTP and GDP bound forms of the small GTPase Ran across the nuclear envelope. Ran shuttles between the cytoplasm and nucleoplasm and cycles between the two nucleotide-bound states depending on accessory proteins in each compartment (Meier 2007; Stewart 2007). Ran GTPase activating protein (RanGAP) is localized on the cytoplasmic side of the NPC, therefore facilitating the conversion of Ran-GTP to Ran-GDP outside the nucleus. As a consequence, the cytoplasm is characterized by high Ran-GDP concentration. The Ran nucleotide exchange factor is sequestered to chromatin and therefore present in the nucleus where it catalyzes the conversion of Ran-GDP to Ran-GTP. The nuclear compartment is therefore characterized by a high Ran-GTP concentration. The Ran-GTP concentration gradient provides the spatial information for nucleocytoplasmic transport. Import complexes consisting of importin α/β and cargo form in the cytoplasm. Upon translocation through the NPC, the binding of Ran-GTP to importin β inside the nucleus causes the dissociation of the import complex, thus releasing the cargo protein (for a recent review, see Stewart 2007). Conversely, export complexes are stabilized by Ran-GTP and dissociate upon hydrolysis to Ran-GDP outside the nucleus (see Merkle 2008).

Most of the components of the Ran cycle have also been identified in plants, with the exception of the Ran nucleotide exchange factor, RCC1, which may have diverged beyond the evolutionary look-back horizon for sequence similarity searches (Meier 2007). RanGAP associates with the NPC in both vertebrates and plants, but utilizes different targeting domains and binding partners at the pore. Vertebrate RanGAP contains a C-terminal targeting domain that binds to Nup358/RanBP2 upon SUMOylation (Mahajan et al. 1998; Matunis et al. 1996). Plant RanGAP utilizes an N-terminal targeting domain termed WPP domain, after a conserved tryptophan-proline-proline motif, that binds to the WIP family of nucleoporins (Jeong et al. 2005; Rose and Meier 2001; Xu et al. 2007a).

Besides facilitating transport through the nuclear pores, the Ran cycle also affects their assembly. A genetic screen for NPC assembly mutants in yeast identified four complementation groups corresponding to different components of the Ran cycle (Ryan et al. 2003). Importin β was found to negatively affect NPC assembly in vertebrates where either increase of Ran-GTP or decrease in importin β triggered the formation of NPC-containing membrane structures (Harel et al. 2003; Walther et al. 2003a,b).

In addition to NPC assembly, the Ran cycle is centrally involved in regulating spindle formation, kinetochore attachment, and nuclear envelope assembly during mitosis (Arnaoutov and Dasso 2005; Ciciarello et al. 2007; Clarke and Zhang 2004). These processes also appear to involve nucleoporins and their interaction with karyopherins and components of the Ran cycle. In vertebrates and yeast, RanGAP and components of the NPC, such as Nup358/RanBP2 of the cytoplasmic fibrils and the entire Nup107-160 complex of the central pore, are relocated during mitosis to the kinetochores (Belgareh et al. 2001; Joseph et al. 2004; Loïodice et al. 2004). Vertebrate Tpr interacts with importin β during nucleocytoplasmic transport (Shah et al. 1998). During mitosis, importin β is relocated to the spindle poles where it regulates Ran-dependent spindle assembly (Ciciarello et al. 2004). The *Drosophila* Tpr homolog Megator appears to be involved in forming a spindle matrix during mitosis (Qi et al. 2004), and its yeast homolog Mlp2p is a component of the spindle pole body (Niepel et al. 2005). So far, the mitotic location of only two plant Nups has been shown. *Arabidopsis* Tpr is located in the vicinity of the spindle, suggestive of a functional similarity to *Drosophila* Megator during mitosis (Xu et al. 2007b). Two members of the WIP family of nucleoporins, WIP1 and WIP2a, colocalize with plant RanGAP at the cell plate in *Arabidopsis* (Xu et al. 2007a), suggestive of plant-specific functions of the Ran cycle and nucleoporins in membrane fusion events during plant cell division.

3.2   *Signal Transduction and Developmental Regulation*

Several mutations in plant nucleoporins have been identified that affect a variety of signaling pathways and developmental processes in plants, often leading to pleiotropic phenotypes. Interestingly, further investigations into the molecular defects in these mutants indicate that multiple molecular mechanisms may be involved in causing the corresponding phenotypes. For example, auxin hormone signaling appears to require protein translocation into the nucleus, while cold stress signaling depends on RNA export, and plant–symbiont interactions involve the generation of a calcium signal at the nuclear envelope. Developmental processes may be affected by micro RNA and SUMO homeostasis.

3.2.1 Hormone Signaling

Mutations in nucleoporins have been found to be involved in the suppression of auxin hormone resistance in *Arabidopsis*. The *auxin-resistant1* (*axr1*) mutation prevents the ubiquitin-mediated degradation of the nuclear Aux/IAA transcriptional repressors of the auxin signaling pathway, thus conferring a constitutive repression of the hormone signal. The gene codes for a subunit of RUB-activating protein, which is the first enzyme in a pathway conjugating the ubiquitin-like protein RUB to cullin proteins. This processing step is required for the formation of a functional SCF complex in the auxin-dependent protein degradation pathway. Two *suppressor of axr1* (*sar*) mutations that restore auxin response in the *axr1* mutant background were found to involve components of the plant NPC (Parry et al. 2006). SAR1 and SAR3 are homologs of vertebrate Nup160 and Nup96, respectively, both components of the conserved symmetrical scaffolding complex critical for nuclear pore assembly in vertebrates (Walther et al. 2003a). Both mutants partially restore the morphological and molecular phenotypes of *axr1* plants with additional shared phenotypes epistatic to the *axr1* phenotype such as early flowering, but are not capable of restoring the decreased levels of RUB-modified cullins (Parry et al. 2006). When crossed with other auxin-resistant mutants, *sar1* and *sar3* are able to partially suppress mutations in other genes of the RUB conjugation pathway, such as the *rce1* mutation affecting a RUB-conjugating E2 enzyme, but *sar1* was found to have no effect on the *aux1-7* mutation in the auxin influx carrier, indicating that the suppression effect is pathway-specific (Cernac et al. 1997; Parry et al. 2006).

Sar1 sar3 double mutant show several defects in nucleocytoplasmic transport on a molecular level. They are deficient in mRNA export and accumulate poly(A)$^+$ RNA in the nucleus, which might contribute to their pleiotropic phenotype, but it appears likely that the *axr1* suppression mechanism is more directly connected to protein import rather than RNA export. A GUS-reporter for the auxin-responsive transcriptional repressor IAA17 was found to be increased in the *sar1* and *sar3* single and double mutants compared to wild-type. However, its nuclear localization in wild-type plants was changed to a distribution throughout the cell in the mutants, indicating that SAR1 and SAR3 nucleoporins are required for its transport or retention in the nucleus (Parry et al. 2006). Since the *axr1* phenotype is due to reduced nuclear degradation of auxin-responsive transcriptional repressors such as IAA17, it appears logical that prevention of the nuclear translocation of these repressors in the *sar1* and *sar3* mutants would constitute a mechanism of suppressing the *axr1* mutation. Since IAA17 degradation is expected to occur in the nucleus in wild-type plants, the reduced translocation rate in *sar1* and *sar3* mutants also provides a possible mechanism for the observed increased protein stability due to lack of degradation (Parry et al. 2006).

However, other mutants impaired in auxin signaling point to RNA export mechanisms as well, suggesting that possibly both types of transport processes may be involved in auxin signaling (also see Merkle 2008). Mutants of the *Arabidopsis* homolog of Tpr/Mlp1p/Mlp2p, components of the nuclear basket in animals, yeast, show similar phenotypes to the *sar1* and *sar3* mutants, impairment of RNA export, and can restore auxin sensitivity in the *axr1* mutant background (Jacob et al. 2007; Xu et al. 2007b). *Arabidopsis attpr* mutant plants show similar reductions in microRNA levels as the *hasty* (*hst*) mutant of the *Arabidopsis* homolog of exportin5, which is also impaired in auxin signaling (Jacob et al. 2007; Park et al. 2005). Most notably, *miR159*, *miR165*, and *miR393* are reduced significantly in the *attpr* mutant. The targets of *miR393* are F-box auxin receptor genes, including transport *inhibitor response1* (*TIR1*), which is part of the functional SCF complex in the auxin-dependent protein degradation pathway and therefore acts in the same pathway as *AXR1*, thus providing a strong link between miRNA metabolism and defects in auxin signaling (Sunkar and Zhu 2004). Global gene expression analysis in a *miR165* overexpressor line also revealed altered expression of genes involved in auxin signaling (Zhou et al. 2007), suggesting that several miRNAs affected in *attpr* may provide a putative link between RNA processing and export and auxin signal transduction.

3.2.2 Cold and Stress Response

Another signal transduction pathway in plants that can be affected by mutations in nucleoporins is the cold stress response. Cold tolerance and response involves the regulation of the expression of CBF cold-responsive transcription factors by the positive regulator ICE1 and the negative regulator HOS1 (Chinnusamy et al. 2003; Ishitani et al. 1998; Lee et al. 2005). A genetic screen designed to identify regulatory components of cold tolerance in transgenic *Arabidopsis* carrying a *CBF3-LUC* reporter gene uncovered a mutation in *nup160*, which is allelic to the auxin resistance suppressor mutation *sar1* (Dong et al. 2006). This mutation affects the expression of the cold-inducible *CBF3-LUC* reporter gene and renders the plants sensitive to chilling and freezing stress. Like *sar1* plants and the *sar1 sar3* double mutant, this independently isolated mutant also showed pleiotropic effects of stunted plant growth and early flowering, accompanied by a defect in mRNA export on the molecular level. However, in contrast to the *sar* mutants affecting the nuclear import of the transcription factor IAA17 (Parry et al. 2006), the transcriptional activator ICE1, the upstream regulator of *CBF3* expression, was not affected in its nuclear import in the cold-sensitive *nup160* mutant (Dong et al. 2006). Therefore, the reduced *CBF3* expression in the mutant is likely caused by a different molecular mechanism rather than a lack of ICE1 translocation to the nucleus.

An intriguing candidate mechanism is the mRNA export defect observed in the mutant. Other mutants with altered cold-stress response were found to also involve proteins of the RNA export machinery. Mutants of the LOS4 DEAD-box RNA helicase phenotypically resemble different nucleoporin mutations defective in RNA export (Gong et al. 2002, 2005). Since the *los4-2* mutation affects a protein implicated in mRNA transport but not a structural component of the nuclear pore, it appears likely that the shared cold response and developmental phenotypes in these mutants are linked directly to the RNA export deficiency rather than a protein import defect (Xu and Meier 2007; also see Merkle 2008).

3.2.3 Plant Innate Immunity and Symbioses

Mutations in nucleoporins and nucleocytoplasmic transport factors have also been found in screens for defects in plant–microbe interactions. Plants have the ability to react to microbes in their biotic environment in several ways. Pathogens trigger the plant defense system while rhizobia or mycorrhizal fungi trigger symbiotic signal transduction leading to the formation of rood nodules or mycorrhiza, respectively.

Plants possess two types of pathogen recognition and defense systems: a nonspecific basal resistance or innate immunity and a specific pathogen response via resistance (R) proteins recognizing pathogens in a gene-for-gene resistance. Both mechanisms lead to the accumulation of reactive oxygen species and the activation of defense genes. The transcriptional reprogramming of plant cells in response to bacterial or fungal pathogen attack can be extensive and involve up to 12% of genes in *Arabidopsis* (Thilmony et al. 2006). While basal resistance confers only a weak response, *R*-gene resistance typically leads to a stronger hypersensitive response and cell death at the infection site (Shen and Schulze-Lefert 2007).

Recent evidence suggests a critical role for nuclear protein import in plant defense mechanisms. A gain-of-function mutation in *Arabidopsis snc1* leads to a constitutively active R protein and therefore chronically activated defense response (Zhang et al. 2003). Three *modifier of snc1* (*mos*) suppressor mutants with restored pathogen susceptibility in the *snc1* mutant background were found to affect components of the NPC and nuclear import machinery. MOS3, which is identical to SAR3, is homologous to the mammalian nucleoporin Nup96 and localizes to the nuclear periphery when fused to GFP (Parry et al. 2006; Zhang and Li 2005). Vertebrate Nup96 and its yeast homolog C-Nup145p serve as components of the conserved symmetrical scaffolding complex critical for nuclear pore assembly (Walther et al. 2003a). MOS6 is importin α3, one of eight *Arabidopsis* importin α homologs, suggesting a requirement for nuclear protein import in plant defense (Palma et al. 2005). MOS7 is homologous to Nup88 and its mutation affects not only SNC1-mediated defense responses, but also basal resistance and systemic acquired resistance (Wiermer et al. 2007). Several candidate proteins have been proposed as targets for MOS3, MOS6, and MOS7 mediated translocation during pathogen defense, including the ENHANCED DISEASE SUSCEPTIBILITY1 (EDS1) protein and its interaction partners PHYTOALEXIN DEFICIENT3 (PAD4) and SENESCENCE ASSOCIATED GENE101 (SAG101), the transcription factor bZIP10, the NONEXPRESSOR OF PATHOGENESIS-RELATED GENES1 (NPR1) protein, and the autoactive snc1 protein itself (Shen and Schulze-Lefert 2007; Wiermer et al. 2007). Interestingly, mice with reduced Nup96 levels are more susceptible to viral infection and impaired in interferon-mediated protein expression (Faria et al. 2006), while the *Drosophila* homolog of Nup88, *members only* (*mbo*), is required for immune response activation after bacterial infection (Uv et al. 2000). Taken together, these findings suggest a functional conservation of the role of certain nucleoporins in immune responses in plants as well as animals.

Plant nucleoporins were also identified in screens for plant–symbiont interaction mutants. Nodulation induced by nitrogen-fixing *Rhizobium* bacteria and colonization of roots by arbuscular mycorrhizal fungi are regulated by a common symbiosis pathway in legumes (Kistner and Parniske 2002). Mutations in several NPC proteins were found to affect both types of plant–microbe interactions in the model legume *Lotus japonicus*. *Nup133* mutants resulted in a temperature-sensitive nodulation deficiency as well as absence of mycorrhizal colonization (Kanamori et al. 2006). *Nup85* mutants similarly showed allele-specific defects in both symbiotic interactions (Saito et al. 2007). In vertebrates, the homologs of these nucleoporins essential for microbial symbiosis in lotus are part of the same complex as the MOS3 homolog Nup96 involved in pathogen response, suggesting a general role of this NPC subcomplex in plant responses to microbes (Wiermer et al. 2007). A common feature of the symbiotic response is Nod-factor-induced calcium spiking. A mutation affecting symbiosis in *Medicago truncatula*, *DMI1* (*Doesn't Make Infections 1*), was found to encode a putative ion channel localized at the nuclear envelope, providing a possible link between ion transport and the nuclear envelope during symbiotic signal transduction (Riely et al. 2007). Both *nup133* and *nup85* mutants lack the calcium-spiking response to Nod-factor, suggesting a role for these nucleoporins in the transduction of the calcium signal at the nuclear envelope. The molecular mechanism of this signal transduction is still unclear. However, activation of calcium channels in the nuclear envelope in *Xenopus* has been shown to lead to structural changes in the conformation and diameter of the NPC, possibly altering its transport capabilities (Erickson et al. 2006).

3.2.4 Flowering Control and Development

Many of the recently identified mutants affecting nucleoporins and components of the nucleocytoplasmic transport machinery in plants have pleiotropic phenotypes, including early flowering, reduced fertility, and characteristic growth defects. *Sar1* mutant plants show a pleiotropic phenotype and are shorter than wild-type with reduced cell division in the root, altered leaf morphology and early flowering (Cernac et al. 1997). The independently identified *nup160* mutant allele of *sar1* was also associated with early flowering and retarded seedling growth (Dong et al. 2006). The phenotype of *sar3* mutants is similar, and the allelic *mos3* mutant was reported to show slightly earlier flowering (Parry et al. 2006; Zhang and Li 2005).

Sar1 sar3 double mutants are more severely affected in development and can be roughly divided into a seedling-lethal group and a non-lethal group of dwarf plants showing very early flowering, changes in phyllotaxy of siliques, and impaired seed production (Parry et al. 2006). Several mutant alleles of *attpr* and *nuclear pore anchor* (*nua*), the *Arabidopsis* homolog of Tpr/Mlp1p/Mlp2p, also show early flowering, stunted growth, reduced fertility and changes in silique arrangements (Jacob et al. 2007; Xu et al. 2007b). While no early flowering phenotype was reported for mutants of the nucleoporins identified from lotus in genetic screens for defects in symbiotic interactions, *nup85* and *nup133*, both mutants had a reduced number of seeds (Kanamori et al. 2006; Saito et al. 2007).

A common theme for many of these mutants is RNA accumulation inside the nucleus, which has been reported for mutant alleles of *sar1/nup160* and *sar3/mos3* and their double mutant, as well as *nua/attpr* (Dong et al. 2006; Jacob et al. 2007; Parry et al. 2006; Xu et al. 2007b). Components of the RNA export machinery have been implicated in similar processes. The mutants of *hasty* (*hst*) and *paused* (*psd*), two exportins of the importin β family implicated in the export of small RNAs such as micro RNAs and tRNAs, show phenotypes similar to nucleoporin mutants including reduced growth, aberrant phyllotaxy and early flowering (Bollmann et al. 2003; Hunter et al. 2003; Li and Chen 2003; Park et al. 2005). *Los4* mutants of a DEAD-box RNA helicase involved in RNA export also phenotypically resemble different nucleoporin mutations with stunted growth and extremely early flowering (Gong et al. 2002, 2005). It is striking that a defect in RNA export appears to only affect certain developmental processes in plants and do so consistently across a wide range of mutants. Recent advances in regulatory RNA research have uncovered the importance of micro RNAs (miRNAs) in developmental processes in plants. Two mutants, *attpr* and *hst*, have been shown to influence miRNA homeostasis, suggesting that interference with miRNA processing and/or transport might be a mechanism common to the observed phenotypes (Jacob et al. 2007; Park et al. 2005).

Changes in miRNA homeostasis have been found in the *attpr* mutant and involve significant reduction of *miR159*, *miR165*, and *miR393* (Jacob et al. 2007). The targets of *miR159* are gibberellic acid-responsive MYB-like transcription factors, MYB33 and MYB65, involved in the regulation of flowering. Overexpression of *miR159* delays flowering in short days by reducing LEAFY transcript levels (Achard et al. 2004), while a double loss-of-function mutation of *miR159a* and *miR159b* has pleiotropic morphological defects with stunted growth, curled leaves, and reduced fertility leading to small siliques and seeds (Allen et al. 2007). The targets of *miRNA165* are homeodomain leucine zipper (HD-ZIP) genes involved in shoot development, and its overexpression causes a pleiotropic phenotype including loss of the shoot apical meristem, altered organ polarity and abnormal carpels, and defects in the development of the vascular tissue (Zhou et al. 2007). As mentioned in the context of auxin signaling, *miR393* targets F-box proteins, including the auxin-dependent protein degradation pathway. It is also involved in the regulation of general stress responses and its expression is strongest in the inflorescence, indicating a possible role in silencing genes in reproductive tissues, and upregulated by cold, dehydration, NaCl, and ABA treatments (Sunkar and Zhu 2004). Bacterial elicitors can lead to an increase in *miR393* in the case of pathogen attack, which contributes to bacterial resistance via repression of auxin signaling (Navarro et al. 2006).

In addition to the changes observed in *attpr*, the *hst* mutant also affects miRNA processing. Changes in *hst* mutant background include a reduction of *miR156*, *miR159*, and *miR171* (Park et al. 2005). The targets of *miR156* are several members of the SQUAMOSA PROMOTER BINDING PROTEIN-LIKE (SPL) family of transcription factors. Constitutive expression of *miR156*-insensitive forms of SLP3, SPL4, and SPL5 was found to promote early flowering, while constitutive expression of *miR156a* delayed flowering (Gandikota et al. 2007; Wu and Poethig 2006). *miR171* was found to be upregulated by stress and targets auxin-responsive SCARECROW-like transcription factors (Sunkar and Zhu 2004).

In summary, many of the phenotypes observed in mutants of nucleoporins or the nucleocytoplasmic transport machinery may be linked to changes in miRNA homeostasis similar to those observed in *attpr* and *hst* plants, thus providing an attractive hypothesis for a general mechanism leading to the observed pleiotropic developmental phenotypes. However, it still remains to be shown whether this link actually exists and whether it is mediated through impaired miRNA export or impaired import of miRNA processing factors into the nucleus. The transport and processing location of miRNAs in plants is so far poorly understood, but a recent study showed that three processing proteins, Dicer-like 1 (DCL1), HYPONASTIC LEAVES1 (HYL1) and SERRATE (SE), colocalized with miRNA primary transcript and cajal body marker proteins in distinct bodies inside *Arabidopsis* nuclei (Fujioka et al. 2007). It is therefore also possible that reduced miRNA processing could be a consequence of a deficiency in protein import in plants.

Another pathway implicated in the early flowering and pleiotropic developmental phenotype is SUMO homeostasis. SUMOylation of nuclear proteins is essential in *Arabidopsis* and has been suggested to play a key role in development and stress response (Saracco et al. 2007). *Nua* mutant alleles of the *Arabidopsis* Tpr homolog phenocopy the *early in short days4* (*esd4*) mutation and both mutations lead to an accumulation of SUMO conjugates (Murtas et al. 2003; Xu et al. 2007b). ESD4 is a SUMO protease concentrated at the nuclear envelope and has been shown to interact with NUA in a yeast two-hybrid assay (Reeves et al. 2002; Xu et al. 2007b).

In the *attpr/nua* and *esd4* mutants, the expression levels of several key regulators of flowering control are affected, providing a link between gene expression changes and phenotype. The transcript level for the floral repressor *FLOWERING LOCUS C* (*FLC*) is decreased, with a concomitant increase in the transcript levels of its downstream targets, the floral activators *FLOWERING LOCUS T* (*FT*) and *SUPPRESSOR OF OVEREXPRESSION OF CO1* (*SOC1*), suggesting that early flowering in these mutants is at least in part mediated through the FLC-dependent autonomous pathway (Jacob et al. 2007; Reeves et al. 2002; Xu et al. 2007b). However, *attpr* and *nua* mutants show a more severe early flowering phenotype compared to the *flc* mutant, indicating that additional mechanisms are likely involved as well (Jacob et al. 2007; Xu et al. 2007b). Winter annuals of *Arabidopsis* contain an epistatic vernalization pathway to delay flowering until spring, which involves the *FRIGIDA* (*FRI*) gene upregulating FLC expression and thus delaying flowering. Interestingly, both *attpr* and the *sar3* nucleoporin mutant are capable of suppressing the late-flowering phenotype of *FRI*, suggesting that a common pathway might exist through which NPC deficiencies might affect flowering time (Jacob et al. 2007).

In summary, many developmental processes are affected in mutants of the NPC and nucleocytoplasmic transport machinery, including flowering time in a wide range of mutants. Gene expression data suggests that the flowering change might be due to a combination of FLC-dependent and FLC-independent mechanisms. Potential links between the NPC and gene expression changes leading to the observed phenotypes might involve RNA export, miRNA processing, and SUMO homeostasis. The available data suggests that likely for each mutation a variety of common processes are involved, affecting a diverse pool of nuclear proteins and thus leading to the observed pleiotropic phenotypes.

4 Evolution of the Nuclear Pore Complex

In the evolution of eukaryotes, the formation of a nuclear envelope surrounding the cellular genome necessitated the evolution of nuclear pores to facilitate the transport across the barrier between the nuclear and cytoplasmic compartments of the cell. Comparative genomics analysis of nuclear envelope and NPC proteins suggests a combination of evolution from prokaryotic precursors, divergence from eukaryotic paralogs, and de novo generation of low complexity (such as repeat amplification) led to the compositions of the nuclear envelope and pore proteomes of modern eukaryotic cells (Mans et al. 2004).

Based on the phylogeny of eukaryotic organisms assuming a "crown group" consisting of plants, fungi, and animals, the last common eukaryotic ancestor would have possessed a core set of nucleoporins and karyopherins and a minimal Ran cycle and ribosomal subunit export system to drive nucleocytoplasmic transport. It has been estimated that these early nuclear pores were already highly complex structures composed of about 20 proteins (Mans et al. 2004). The nuclear import system in these early eukaryotes likely included the karyopherins importin α and β, which have additional functions in spindle formation and nuclear envelope and pore complex assembly, respectively (Askjaer et al. 2002; Harel et al. 2003a; Zhang et al. 2002). Subsequent evolution then involved multiple duplications of nucleoporins and karyopherins as well as recruitment of novel nucleoporins through domain assembly or late horizontal gene transfer from bacteria (Mans et al. 2004).

The NPC shares a number of protein domains with the COPII vesicle coat system of the endoplasmic reticulum. Most notably, components of the Sec13p-Sec31p COPII subcomplex are also part of the NPC (Enninga et al. 2003; Siniossoglou et al. 1996), while karyopherins share HEAT domains closely related to proteins involved in vesicle formation at the ER (Mans et al. 2004). In fact, the central core scaffold of the NPC is structurally analogous to vesicle-coating complexes and utilizes a surprisingly simple architecture of only two different domain folds to coat the surface of the curved pore membrane (Alber et al. 2007). This apparent link between the NPC and cytoplasmic vesicle biogenesis and coating complexes supports the hypothesis that the nuclear envelope and NPC likely evolved from or co-evolved with the ER and the use of phagotrophy in the early eukaryote (Cavalier-Smith 2002). The architecture of modern eukaryotic NPCs appears highly modular and consists of repeating structural units that might have formed through gene duplication and divergence events. Thus, the primordial NPC might have consisted of a much simpler structure containing only a minimum of modules, which subsequently duplicated and expanded during the evolution of eukaryotes, giving rise to the more complex NPCs of modern eukaryotes (Alber et al. 2007).

A pivotal event in the evolution of the nuclear envelope and pores likely was the radiation of the Ras family of signaling GTPases leading to the development of the Ran cycle (Mans et al. 2004). Not only is the Ran cycle required for nucleocytoplasmic transport through the pore, but also for the assembly of the nuclear envelope and NPC (Ryan et al. 2003; Walther et al. 2003b). Ran-GTP recruits vesicles to chromatin and promotes their fusion to form the double membrane of the nuclear envelope (Clarke and Zhang 2004). An interesting example of convergent evolution by domain tinkering can be observed in the Ran GTPase activating protein (RanGAP). RanGAP is attached to the nuclear pore in both higher plants and vertebrates, but not in yeast, via different targeting domains and interaction partners at the NPC. The plant and animal RanGAP targeting domains are attached to opposite termini of the protein and are not functionally interchangeable (Jeong et al. 2005). Targeting of mammalian RanGAP requires SUMOylation of its C-terminal targeting domain to associate with Nup358, an animal-specific nucleoporin (Mahajan et al. 1998; Matunis et al. 1996). On the other hand, targeting of plant RanGAP requires its N-terminal domain and interaction with the plant-specific WIP family of nucleoporins (Xu et al. 2007a). This suggests that the targeting of RanGAP to the NPC evolved independently after the divergence of the major eukaryotic kingdoms, likely as an adaptation to the evolution of open mitosis in the plant and animal kingdoms in which the nuclear envelope disassembles (Rose and Meier 2001).

5 Conclusions and Outlook

While our understanding of the molecular composition of the plant nuclear pore is still in its beginning stages, the recent emergence of nucleoporins and components of the nucleocytoplasmic transport machinery as developmental and signal transduction mutants is providing a tantalizing glimpse into the involvement of this structure in a variety of processes in plant cells. The fact that these mutations prove viable in plants provides valuable tools for the future study of not only plant nuclear pore structure and function, but by inference possibly yet unknown functions of the homologous vertebrate and yeast proteins. Future studies may reveal additional plant-specific nucleoporins as well as further pinpoint the pore activities required to control developmental and signal transduction pathways in plant cells.

References

Ach RA, Gruissem W (1994) A small nuclear GTP-binding protein from tomato suppresses a *Schizosaccharomyces pombe* cell-cycle mutant. Proc Natl Acad Sci USA 91:5863–5867

Achard P, Herr A, Baulcombe DC, Harberd NP (2004) Modulation of floral development by a gibberellin-regulated microRNA. Development 131:3357–3365

Alber F, Dokudovskaya S, Veenhoff LM, Zhang W, Kipper J, Devos D, Suprapto A, Karni-Schmidt O, Williams R, Chait BT, Sali A, Rout MP (2007) The molecular architecture of the nuclear pore complex. Nature 450:695–701

Allen RS, Li J, Stahle MI, Dubroué A, Gubler F, Millar AA (2007) Genetic analysis reveals functional redundancy and the major target genes of the *Arabidopsis* miR159 family. Proc Natl Acad Sci USA 104:16371–16376

Arnaoutov A, Dasso M (2005) Ran-GTP regulates kinetochore attachment in somatic cells. Cell Cycle 4:1161–1165

Askjaer P, Galy V, Hannak E, Mattaj IW (2002) Ran GTPase cycle and importins α and β are essential for spindle formation and nuclear envelope assembly in living *Caenorhabditis elegans* embryos. Mol Biol Cell 13:4355–4370

Ballas N, Citovsky V (1997) Nuclear localization signal binding protein from *Arabidopsis* mediates nuclear import of *Agrobacterium* VirD2 protein. Proc Natl Acad Sci USA 94:10723–10728

Bayliss R, Ribbeck K, Akin D, Kent HM, Feldherr CM, Görlich D, Stewart M (1999) Interaction between NTF2 and xFxFG-containing nucleoporins is required to mediate nuclear import of RanGDP. J Mol Biol 293:579–593

Belgareh N, Rabut G, Baï SW, van Overbeek M, Beaudouin J, Daigle N, Zatsepina OV, Pasteau F, Labas V, Fromont-Racine M, Ellenberg J, Doye V (2001) An evolutionarily conserved NPC subcomplex, which redistributes in part to kinetochores in mammalian cells. J Cell Biol 154:1147–1160

Ben-Efraim I, Gerace L (2001) Gradient of increasing affinity of importin β for nucleoporins along the pathway of nuclear import. J Cell Biol 152:411–417

Bollman KM, Aukerman MJ, Park MY, Hunter C, Berardini TZ, Poethig RS (2003) HASTY, the *Arabidopsis* ortholog of exportin 5/MSN5, regulates phase change and morphogenesis. Development 130:1493–1504

Callan HG, Tomlin SG (1950) Experimental studies on amphibian oocyte nuclei. I. Investigations of the structure of the nuclear membrane by means of the electron microscope. Proc R Soc London B 137:367–378

Cavalier-Smith T (2002) The phagotrophic origin of eukaryotes and phylogenetic classification of Protozoa. Int J Syst Evol Microbiol 52:297–354

Cernac A, Lincoln C, Lammer D, Estelle M (1997) The *SAR1* gene of *Arabidopsis* acts downstream of the *AXR1* gene in auxin response. Development 124:1583–1591

Chial HJ, Rout MP, Giddings TH, Winey M (1998) *Saccharomyces cerevisiae* Ndc1p is a shared component of nuclear pore complexes and spindle pole bodies. J Cell Biol 143:1789–1800

Chinnusamy V, Ohta M, Kanrar S, Lee BH, Hong X, Agarwal M, Zhu JK (2003) ICE1: a regulator of cold-induced transcriptome and freezing tolerance in *Arabidopsis*. Genes Dev 17:1043–1054

Ciciarello M, Mangiacasale R, Thibier C, Guarguaglini G, Marchetti E, Fiore1 BD, Lavia P (2004) Importin β is transported to spindle poles during mitosis and regulates Ran-dependent spindle assembly factors in mammalian cells. J Cell Sci 117:6511–6522

Ciciarello M, Mangiacasale R, Lavia P (2007) Spatial control of mitosis by the GTPase Ran. Cell Mol Life Sci 64:1891–1914

Clarke PR, Zhang C (2004) Spatial and temporal control of nuclear envelope assembly by Ran GTPase. Symp Soc Exp Biol 56:193–204

Cohen M, Wilson KL, Gruenbaum Y (2001) Membrane proteins of the nuclear pore complex: Gp210 is conserved in *Drosophila*, *C. elegans* and *A. thaliana*. Gene Ther Mol Biol 6:47–55

Cordes VC, Reidenbach S, Rackwitz HR, Franke WW (1997) Identification of protein p270/Tpr as a constitutive component of the nuclear pore complex-attached intranuclear filaments. J Cell Biol 136:515–529

Cronshaw JM, Krutchinsky AN, Zhang W, Chait BT, Matunis MJ (2002) Proteomic analysis of the mammalian nuclear pore complex. J Cell Biol 158:915–927

Denning DP, Patel SS, Uversky V, Fink AL, Rexach M (2003) Disorder in the nuclear pore complex: The FG repeat regions of nucleoporins are natively unfolded. Proc Natl Acad Sci USA 100:2450–2455

Dong CH, Hu X, Tang W, Zheng X, Kim YS, Lee BH, Zhu JK (2006) A putative *Arabidopsis* nucleoporin, AtNUP160, is critical for RNA export and required for plant tolerance to cold stress. Mol Cell Biol 26:9533–9543

Dreger M, Otto H (2004) The nuclear envelope proteome. Symp Soc Exp Biol 56:9–40

Drummond S, Allen T (2004) Structure, function and assembly of the nuclear pore complex. Symp Soc Exp Biol 56:89–114

Enninga J, Levay A, Fontoura BM (2003) Sec13 shuttles between the nucleus and the cytoplasm and stably interacts with Nup96 at the nuclear pore complex. Mol Cell Biol 23:7271–7284

Erickson ES, Mooren OL, Moore D, Krogmeier JR, Dunn RC (2006) The role of nuclear envelope calcium in modifying nuclear pore complex structure. Can J Physiol Pharmacol 84:309–318

Evans DE, Bryant JA, Hutchison C (2004) The nuclear envelope: a comparative overview. Symp Soc Exp Biol 56:1–8

Faria AM, Levay A, Wang Y, Kamphorst AO, Rosa ML, Nussenzveig DR, Balkan W, Chook YM, Levy DE, Fontoura BM (2006) The nucleoporin Nup96 is required for proper expression of interferon-regulated proteins and functions. Immunity 24:295–304

Favreau C, Worman HJ, Wozniak RW, Frappier T, Courvalin JC (1996) Cell cycle-dependent phosphorylation of nucleoporins and nuclear pore membrane protein Gp210. Biochemistry 35:8035–8044

Feldherr CM, Akin D (1990) EM visualization of nucleocytoplasmic transport processes. Electron Microsc Rev 3:73–86

Fontoura BMA, Blobel G, Matunis MJ (1999) A conserved biogenesis pathway for nucleoporins: proteolytic processing of a 186-kilodalton precursor generates Nup98 and the novel nucleoporin, Nup96. J Cell Biol 144:1097–1112

Fribourg S, Braun IC, Izaurralde E, Conti E (2001) Structural basis for the recognition of a nucleoporin FG repeat by the NTF2-like domain of the TAP/p15 mRNA nuclear export factor. Mol Cell 8:645–656

Fujioka Y, Utsumi M, Ohba Y, Watanabe Y (2007) Location of a possible miRNA processing site in SmD3/SmB nuclear bodies in *Arabidopsis*. Plant Cell Physiol 48:1243–1253

Gandikota M, Birkenbihl RP, Höhmann S, Cardon GH, Saedler H, Huijser P (2007) The miRNA156/157 recognition element in the 3′ UTR of the *Arabidopsis* SBP box gene SPL3 prevents early flowering by translation inhibition in seedlings. Plant J 49:683–693

Gasiorowski JZ, Dean DA (2003) Mechanisms of nuclear transport and interventions. Adv Drug Deliv Rev 55:703–716

Gerace L, Burke B (1988) Functional organization of the nuclear envelope. Annu Rev Cell Biol 4:335–774

Gerace L, Foisner R (1994) Integral membrane proteins and dynamic organization of the nuclear envelope. Trends Cell Biol 4:127–131

Goldberg M, Harel A, Gruenbaum Y (1999) The nuclear lamina: molecular organization and interaction with chromatin. Crit Rev Eukaryot Gene Expr 9:285–293

Gong Z, Lee H, Xiong L, Jagendorf A, Stevenson B, Zhu JK (2002) RNA helicase-like protein as an early regulator of transcription factors for plant chilling and freezing tolerance. Proc Natl Acad Sci USA 99:11507–11512

Gong Z, Dong CH, Lee H, Zhu J, Xiong L, Gong D, Stevenson B, Zhu JK (2005) A DEAD box RNA helicase is essential for mRNA export and important for development and stress responses in *Arabidopsis*. Plant Cell 17:256–267

Görlich D, Kutay U (1999) Transport between the cell nucleus and the cytoplasm. Annu Rev Cell Dev Biol 15:607–660

Haltiwanger RS, Grove K, Philipsberg GA (1998) Modulation of *O*-linked *N*-acetylglucosamine levels on nuclear and cytoplasmic proteins *in vivo* using the peptide *O*-GlcNAc-β - *N*-acetylglucosaminidase inhibitor *O*-(2-acetamido-2-deoxy-ᴅ-glucopyranosylidene)amino-*N*-phenylcarbamate. J Biol Chem 273:3611–3617

Hang J, Dasso M (2002) Association of the human SUMO-1 protease SENP2 with the nuclear pore. J Biol Chem 277:19961–19966

Harel A, Chan RC, Lachish-Zalait A, Zimmerman E, Elbaum M, Forbes DJ (2003a) Importin β negatively regulates nuclear membrane fusion and nuclear pore complex assembly. Mol Biol Cell 14:4387–4396

Harel A, Orjalo AV, Vincent T, Lachish-Zalait A, Vasu S, Shah S, Zimmermann E, Elbaum M, Forbes DJ (2003b) Removal of a single pore subcomplex results in vertebrate nuclei devoid of nuclear pores. Mol Cell 11:853–864

Heese-Peck A, Raikhel NV (1998) A glycoprotein modified with terminal *N*-acetylglucosamine and localized at the nuclear rim shows sequence similarity to aldose-1-epimerases. Plant Cell 10:599–612

Heese-Peck A, Cole RN, Borkhsenious ON, Hart GW, Raikhel NV (1995) Plant nuclear pore complex proteins are modified by novel oligosaccharides with terminal *N*-acetylglucosamine. Plant Cell 7:1459–1471

Hicks GR, Smith HM, Lobreaux S, Raikhel NV (1996) Nuclear import in permeabilized protoplasts from higher plants has unique features. Plant Cell 8:1337–1352

Hinshaw JE, Carragher BO, Milligan RA (1992) Architecture and design of the nuclear pore complex. Cell 69:1133–1141

Hübner S, Smith HMS, Hu W, Chan CK, Rihs HP, Paschal BM, Raikhel NV, Jans DA (1999) Plant importin α binds nuclear localization sequences with high affinity and can mediate nuclear import independent of importin β. J Biol Chem 274:22610–22617

Hunter CA, Aukerman MJ, Sun H, Fokina M, Poethig RS (2003) *PAUSED* encodes the *Arabidopsis* exportin-t ortholog. Plant Physiol 132:2135–2143

Ishitani M, Xiong L, Lee H, Stevenson B, Zhu JK (1998) *HOS1*, a genetic locus involved in cold-responsive gene expression in *Arabidopsis*. Plant Cell 10:1151–1161

Jacob Y, Mongkolsiriwatana C, Veley KM, Kim SY, Michaels SD (2007) The nuclear pore protein AtTPR is required for RNA homeostasis, flowering time, and auxin signaling. Plant Physiol 144:1383–1390

Jeong SY, Rose A, Joseph J, Dasso M, Meier I (2005) Plant-specific mitotic targeting of RanGAP requires a functional WPP domain. Plant J 42:270–282

Joseph J, Liu ST, Jablonski SA, Yen TJ, Dasso M (2004) The RanGAP1-RanBP2 complex is essential for microtubule-kinetochore interactions *in vivo*. Curr Biol 14:611–617

Kanamori N, Madsen LH, Radutoiu S, Frantescu M, Quistgaard EMH, Miwa H, Downie JA, James EK, Felle HH, Haaning LL, Jensen TH, Sato S, Nakamura Y, Tabata S, Sandal N, Stougaard J (2006) A nucleoporin is required for induction of Ca^{2+} spiking in legume nodule development and essential for rhizobial and fungal symbiosis. Proc Natl Acad Sci USA 103:359–364

Kiger AA, Gigliotti S, Fuller MT (1999) Developmental genetics of the essential *Drosophila* nucleoporin *nup154*: allelic differences due to an outward-directed promoter in the *P*-element 3′ end. Genetics 153:799–812

Kistner C, Parniske M (2002) Evolution of signal transduction in intracellular symbiosis. Trends Plant Sci 7:511–518

Krull S, Thyberg J, Björkroth B, Rackwitz HR, Cordes VC (2004) Nucleoporins as components of the nuclear pore complex core structure and Tpr as the architectural element of the nuclear basket. Mol Biol Cell 15:4261–4277

Lee BH, Henderson DA, Zhua JK (2005) The *Arabidopsis* cold-responsive transcriptome and its regulation by ICE1. Plant Cell 17:3155–3175

Li J, Chen X (2003) *PAUSED*, a putative exportin-at, acts pleiotropically in *Arabidopsis* development but is dispensable for viability. Plant Physiol 132:1913–1924

Li D, Roberts R (2001) WD-repeat proteins: structure characteristics, biological function, and their involvement in human diseases. Cell Mol Life Sci 58:2085–2097

Lim RY, Aebi U, Fahrenkrog B (2008) Towards reconciling structure and function in the nuclear pore complex. Histochem Cell Biol 129:105–116

Loïodice I, Alves A, Rabut G, Van Overbeek M, Ellenberg J, Sibarita JB, Doye V (2004) The entire Nup107-160 complex, including three new members, is targeted as one entity to kinetochores in mitosis. Mol Biol Cell 15:3333–3344

Macara IG (2001) Transport into and out of the nucleus. Microbiol Mol Biol Rev 65:570–594

Macaulay C, Meier E, Forbes DJ (1995) Differential mitotic phosphorylation of proteins of the nuclear pore complex. J Biol Chem 270:254–262

Mahajan R, Delphin C, Guan T, Gerace L, Melchior F (1997) A small ubiquitin-related polypeptide involved in targeting RanGAP1 to nuclear pore complex protein RanBP2. Cell 88:97–107

Mahajan R, Gerace L, Melchior F (1998) Molecular characterization of the SUMO-1 modification of RanGAP1 and its role in nuclear envelope association. J Cell Biol 140:259–270

Mans BJ, Anantharaman V, Aravind L, Koonin EV (2004) Comparative genomics, evolution and origins of the nuclear envelope and nuclear pore complex. Cell Cycle 3:1612–1637

Mansfeld J, Güttinger S, Hawryluk-Gara LA, Panté N, Mall M, Galy V, Haselmann U, Mühlhäusser P, Wozniak RW, Mattaj IW, Kutay U, Antonin W (2006) The conserved transmembrane nucleoporin NDC1 is required for nuclear pore complex assembly in vertebrate cells. Mol Cell 22:93–103

Mattaj IW, Englmeier L (1998) Nucleocytoplasmic transport: the soluble phase. Annu Rev Biochem 67:265–306

Matunis MJ, Coutavas E, Blobel G (1996) A novel ubiquitin-like modification modulates the partitioning of the Ran-GTPase-activating protein RanGAP1 between the cytosol and the nuclear pore complex. J Cell Biol 135:1457–1470

Meier I (2007) Composition of the plant nuclear envelope: theme and variations. J Exp Bot 58:27–34

Merkle T (2001) Nuclear import and export of proteins in plants: a tool for the regulation of signalling. Planta 213:499–517

Merkle T (2008) Nuclear export of proteins and RNA. Plant Cell Monogr., doi:10.1007/7089_2008_25

Merkle T, Haizel T, Matsumoto T, Harter K, Dallmann G, Nagy F (1994) Phenotype of the fission yeast cell cycle regulatory mutant *pim1-46* is suppressed by a tobacco cDNA encoding a small, Ran-like GTP-binding protein. Plant J 6:555–565

Merkle T, Leclerc D, Marshallsay C, Nagy F (1996) A plant *in vitro* system for the nuclear import of proteins. Plant J 10:1177–1186

Miller BR, Powers M, Park M, Fischer W, Forbes DJ (2000) Identification of a new vertebrate nucleoporin, Nup188, with the use of a novel organelle trap assay. Mol Biol Cell 11:3381–3396

Murtas G, Reeves PH, Fu YF, Bancroft I, Dean C, Coupland G (2003) A nuclear protease required for flowering-time regulation in *Arabidopsis* reduces the abundance of SMALL UBIQUITIN-RELATED MODIFIER conjugates. Plant Cell 17:705–721

Navarro L, Dunoyer P, Jay F, Arnold B, Dharmasiri N, Estelle M, Voinnet O, Jones JDG (2006) A plant miRNA contributes to antibacterial resistance by repressing auxin signaling. Science 312:436–439

Niepel M, Strambio-de-Castillia C, Fasolo J, Chait BT, Rout MP (2005) The nuclear pore complex-associated protein, Mlp2p, binds to the yeast spindle pole body and promotes its efficient assembly. J Cell Biol 70:225–235

Paddy MR (1998) The Tpr protein: linking structure and function in the nuclear interior? Am J Hum Genet 63:305–310

Palma K, Zhang Y, Li X (2005) An importin α homolog, MOS6, plays an important role in plant innate immunity. Curr Biol 15:1129–1135

Panté N, Aebi U (1996) Sequential binding of import ligands to distinct nucleopore regions during their nuclear import. Science 273:1729–1732

Panté N, Kann M (2002) Nuclear pore complex is able to transport macromolecules with diameters of ~39 nm. Mol Biol Cell 13:425–434

Park MY, Wu G, Gonzalez-Sulser A, Vaucheret H, Poethig RS (2005) Nuclear processing and export of microRNAs in *Arabidopsis*. Proc Natl Acad Sci USA 102:3691–3696

Parry G, Ward S, Cernac A, Dharmasiri S, Estelle M (2006) The *Arabidopsis* SUPPRESSOR OF AUXIN RESISTANCE proteins are nucleoporins with an important role in hormone signaling and development. Plant Cell 18:1590–1603

Pay A, Resch K, Frohnmeyer H, Fejes E, Nagy F, Nick P (2002) Plant RanGAPs are localized at the nuclear envelope in interphase and associated with microtubules in mitotic cells. Plant J 30:699–709

Pemberton LF, Paschal BM (2005) Mechanisms of receptor-mediated nuclear import and nuclear export. Traffic 6:187–198

Qi H, Rath U, Wang D, Xu YZ, Ding Y, Zhang W, Blacketer MJ, Paddy MR, Girton J, Johansen J, Johansen KM (2004) Megator, an essential coiled-coil protein that localizes to the putative spindle matrix during mitosis in Drosophila. Mol Biol Cell 15:4854–4865

Reeves PH, Murtas G, Dash S, Coupland G (2002) *early in short days 4*, a mutation in *Arabidopsis* that causes early flowering and reduces the mRNA abundance of the floral repressor *FLC*. Development 129:5349–5361

Reichelt R, Holzenburg A, Buhle Jr EL, Jarnik M, Engel A, Aebi U (1990) Correlation between structure and mass distribution of the nuclear pore complex and of distinct pore complex components. J Cell Biol 110:883–894

Riely BK, Lougnon G, Ané JM, Cook DR (2007) The symbiotic ion channel homolog DMI1 is localized in the nuclear membrane of *Medicago truncatula* roots. Plant J 49:208–216

Ris H (1997) High-resolution field-emission scanning electron microscopy of nuclear pore complex. Scanning 19:368–375

Roberts K, Northcote DH (1970) Structure of the nuclear pore in higher plants. Nature 228:385–386

Rose A, Meier I (2001) A domain unique to plant RanGAP is responsible for its targeting to the plant nuclear rim. Proc Natl Acad Sci USA 98:15377–15382

Rose A, Patel S, Meier I (2004) Plant nuclear envelope proteins. Symp Soc Exp Biol 56:69–88

Rosenblum JS, Blobel G (1999) Autoproteolysis in nucleoporin biogenesis. Proc Natl Acad Sci USA 96:11370–11375

Rout MP, Aitchison JD, Suprapto A, Hjertaas K, Zhao Y, Chait BT (2000) The yeast nuclear pore complex: composition, architecture, and transport mechanism. J Cell Biol 148:635–651

Ryan KJ, McCaffery JM, Wente SR (2003) The Ran GTPase cycle is required for yeast nuclear pore complex assembly. J Cell Biol 160:1041–1053

Saito K, Yoshikawa M, Yano K, Miwa H, Uchida H, Asamizu E, Sato S, Tabata S, Imaizumi-Anraku H, Umehara Y, Kouchi H, Murooka Y, Szczyglowski K, Downie JA, Parniske M, Hayashi M, Kawaguchi M (2007) NUCLEOPORIN85 is required for calcium spiking, fungal and bacterial symbioses, and seed production in *Lotus japonicus*. Plant Cell 19:610–624

Saracco SA, Miller MJ, Kurepa J, Vierstra RD (2007) Genetic analysis of SUMOylation in *Arabidopsis*: Conjugation of SUMO1 and SUMO2 to nuclear proteins is essential. Plant Physiol 145:119–134

Shah S, Tugendreich S, Forbes D (1998) Major binding sites for the nuclear import receptor are the internal nucleoporin Nup153 and the adjacent nuclear filament protein Tpr. J Cell Biol 141:31–49

Shen QH, Schulze-Lefert P (2007) Rumble in the nuclear jungle: compartmentalization, trafficking, and nuclear action of plant immune receptors. EMBO J 26:4293–4301

Siniossoglou S, Wimmer S, Rieger M, Doye V, Tekotte H, Weise C, Emig S, Segref A, Hurt EC (1996) A novel complex of nucleoporins, which includes Sec13p and a Sec13p homolog, is essential for normal nuclear pores. Cell 26:265–275

Smith HMS, Raikhel NV (1999) Protein targeting to the nuclear pore. What can we learn from plants? Plant Physiol 119:1157–1163

Smith HMS, Hicks GR, Raikhel NV (1997) Importin α from *Arabidopsis thaliana* is a nuclear import receptor that recognizes three classes of import signals. Plant Physiol 114:411–417

Stewart M (2007) Molecular mechanism of the nuclear protein import cycle. Nat Rev Mol Cell Biol 8:195–208

Stoffler D, Fahrenkrog B, Aebi U (1999) The nuclear pore complex: from molecular architecture to functional dynamics. Curr Opin Cell Biol 11:391–401

Strässer K, Bassler J, Hurt E (2000) Binding of the Mex67p/Mtr2p heterodimer to FXFG, GLFG, and FG repeat nucleoporins is essential for nuclear mRNA export. J Cell Biol 150:695–706

Stavru F, Hülsmann BB, Spang A, Hartmann E, Cordes VC, Görlich D (2006) NDC1: a crucial membrane-integral nucleoporin of metazoan nuclear pore complexes. J Cell Biol 173:509–519

Sunkar R, Zhu JK (2004) Novel and stress-regulated microRNAs and other small RNAs from *Arabidopsis*. Plant Cell 16:2001–2019

Teixeira MT, Siniossoglou S, Podtelejnikov S, Bénichou JC, Mann M, Dujon B, Hurt E, Fabre E (1997) Two functionally distinct domains generated by *in vivo* cleavage of Nup145p: a novel biogenesis pathway for nucleoporins. EMBO J 16:5086–5097

Thilmony R, Underwood W, He SY (2006) Genome-wide transcriptional analysis of the *Arabidopsis thaliana* interaction with the plant pathogen *Pseudomonas syringae* pv. tomato DC3000 and the human pathogen *Escherichia coli* O157:H7. Plant J 46:34–53

Tzafrir I, Dickerman A, Brazhnik O, Nguyen Q, McElver J, Frye C, Patton D, Meinke D (2003) The *Arabidopsis* SeedGenes Project. *Nucleic Acids Res* 31:90–93

Uv AE, Roth P, Xylourgidis N, Wickberg A, Cantera R, Samakovlis C (2000) *members only* encodes a *Drosophila* nucleoporin required for rel protein import and immune response activation. Genes Dev 14:1945–1957

van Nocker S, Ludwig P (2003) The WD-repeat protein superfamily in *Arabidopsis*: conservation and divergence in structure and function. BMC Genomics 4:50

Walther TC, Alves A, Pickersgill H, Loïodice I, Hetzer M, Galy V, Hülsmann BB, Köcher T, Wilm M, Allen T, Mattaj IW, Doye V (2003a) The conserved Nup107-160 complex is critical for nuclear pore complex assembly. Cell 113:195–206

Walther TC, Askajer P, Gentzel M, Habermann A, Griffiths G, Wilm M, Mattaj JW, Hetzer M (2003b) RanGTP mediates nuclear pore complex assembly. Nature 424:689–694

Wiermer M, Palma K, Zhang Y, Li X (2007) Should I stay or should I go? Nucleocytoplasmic trafficking in plant innate immunity. Cell Microbiol 9:1880–1890

Wu G, Poethig RC (2006) Temporal regulation of shoot development in *Arabidopsis thaliana* by *miR156* and its target *SPL3*. Development 133:3539–3547

Xu XM, Meier I (2007) The nuclear pore comes to the fore. Trends Plant Sci 13:20–27

Xu XM, Meulia T, Meier I (2007a) Anchorage of plant RanGAP to the nuclear envelope involves novel nuclear-pore-associated proteins. Curr Biol 17:1157–1163

Xu XM, Rose A, Muthuswamy S, Jeong SY, Venkatakrishnan S, Zhao Q, Meier I (2007b) NUCLEAR PORE ANCHOR, the *Arabidopsis* homolog of Tpr/Mlp1/Mlp2/Megator, is involved in mRNA export and SUMO homeostasis and affects diverse aspects of plant development. Plant Cell 19:1537–1548

Yamamoto N, Deng XW (1999) Protein nucleocytoplasmic transport and its light regulation in plants. Genes Cells 4:489–500

Zhang Y, Li X (2005) A putative nucleoporin 96 is required for both basal defense and constitutive resistance responses mediated by *suppressor of npr1–1, constitutive 1*. Plant Cell 17:1306–1316

Zhang Q, Skepper JN, Yang F, Davies JD, Hegyi L, Roberts RG, Weissberg PL, Ellis JA, Shanahan CM (2001) Nesprins: a novel family of spectrin-repeat-containing proteins that localize to the nuclear membrane in multiple tissues. J Cell Sci 114:4485–4498

Zhang C, Hutchins JR, Mühlhäusser P, Kutay U, Clarke PR (2002) Role of importin β in the control of nuclear envelope assembly by Ran. Curr Biol 12:498–502

Zhang Y, Goritschnig S, Dong X, Li X (2003) A gain-of-function mutation in a plant disease resistance gene leads to constitutive activation of downstream signal transduction pathways in *suppressor of npr1-1, constitutive 1*. Plant Cell 15:2636–2646

Zhao X, Wu CY, Blobel G (2004) Mlp-dependent anchorage and stabilization of a desumoylating enzyme is required to prevent clonal lethality. J Cell Biol 167:605–611

Zhou GK, Kubo M, Zhong R, Demura T, Ye ZH (2007) Overexpression of miR165 affects apical meristem formation, organ polarity establishment and vascular development in *Arabidopsis*. Plant Cell Physiol 48:391–404

Zimowska G, Aris JP, Paddy MR (1997) A *Drosophila* Tpr protein homolog is localized both in the extrachromosomal channel network and to nuclear pore complexes. J Cell Sci 110:927–944

Nuclear Export of Proteins and RNA

Thomas Merkle

Abstract Several nuclear export receptors that facilitate the export of proteins and small RNAs from the nucleus to the cytoplasm have been functionally characterized in *Arabidopsis thaliana* in the past few years. With the specific cargo molecules they transport, the export receptors supply the cytoplasm with information, resulting in changes in cellular events. In this way, nuclear export receptors contribute to signal transduction that is highlighted in mutants with impaired export activity. In addition, nuclear export of proteins is part of a control layer for gene expression that is unique to eukaryotes, namely nucleocytoplasmic partitioning of regulatory proteins. This is revealed by growing evidence for shuttling proteins that participate in different signaling cascades, including phytohormone signaling. Although a plant mRNA export receptor is still missing in plants, several genes encoding protein factors that function in mRNA export to the cytoplasm were characterized by their mutant phenotypes. These findings show that cold, stress, or phytohormone signaling cascades and the control of flowering are especially sensitive to impaired mRNA biogenesis and export.

1 Introduction

Fast and efficient transport of molecules and macromolecular complexes across the nuclear envelope via the nuclear pore complexes (NPCs) is required to exchange materials and information between the nucleus and the cytoplasm in a living cell. Not long ago nucleo-cytoplasmic transport was largely regarded as a necessary consequence of the spatial separation of translation in the cytoplasm and transcription

T. Merkle
Institute of Genome Research and Systems Biology, Faculty of Biology III,
University of Bielefeld, 33594 Bielefeld, Germany
e-mail: tmerkle@cebitec.uni-bielefeld.de

Plant Cell Monogr, doi:10.1007/7089_2008_25

in the nucleus, only. Thus, karyophilic proteins have to be imported into and RNAs need to be exported from the nucleus to allow proper function of cellular processes. However, this simplified picture has changed dramatically. As for nuclear import, nuclear export provides numerous points for control and regulation.

In this chapter, I will first introduce the nuclear export machinery in general and compare it with the current knowledge in plants. Second, I will discuss the importance of nuclear export of proteins for regulation of specific cellular processes and for signaling. Then, the impact of nuclear export of small RNAs and of mRNA on regulation of growth and development will be exemplified by specific plant mutants, complemented by the discussion of proteins that may be implicated in RNA export in plants, based on their homology with yeast or vertebrate proteins.

2   Nuclear Export Receptors

2.1   Two Receptor Classes

In general, there are two different classes of nuclear export receptors. Members of the first class are characterized by their ability to interact with the small GTPase Ran, which is why they are also referred to as Ran-binding proteins (RanBPs). In contrast, members of the second class do not bind to Ran and are structurally unrelated to the members of the first class of proteins.

Nuclear transport receptors that interact with Ran form a family of related proteins that was named after their founding member Importin beta (or Karyopherin beta). Depending on whether they facilitate cargo import or export, members of this family are called importins or transportins and exportins, respectively (reviewed in: Görlich and Kutay 1999; Merkle 2003; Pemberton and Paschal 2005) (Fig. 1). In the *Arabidopsis* genome, 17 genes encode Importin beta-like nuclear transport receptors (Fig. 2), whereas the human genome contains at least 21 genes. In interphase, Ran regulates directionality of nuclear transport processes that depend on Importin beta-like receptors. In other words, Importin beta-like receptors make use of positional information that is created by the Ran GTPase system in the cell. The GTPase cycle of Ran is distributed between the nucleus and the cytoplasm because its regulatory proteins show distinct localizations within the cell. While Ran itself shuttles between the nucleus and the cytoplasm, the nucleotide exchange factor that catalyzes GTP loading of Ran (Regulator of Chromosome Condensation 1, RCC1 in humans) is associated with chromatin. Thus, a high concentration of the GTP-bound form of Ran (RanGTP) marks the position of the chromosomes in the cell, which corresponds to the nuclear compartment in interphase. Exportins are characterized by a high affinity to their cargo substrates in the presence of RanGTP that becomes a part of the export complex that forms in the nucleus. On the other hand, localization of the proteins that act together to catalyze GTP hydrolysis on Ran, namely Ran-binding protein 1 (RanBP1) and the GTPase-activating protein for Ran

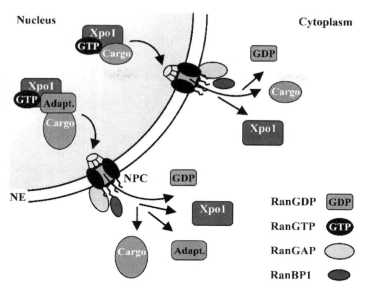

Fig. 1 Nuclear export by Importin beta-like receptors. The RanGTP gradient over the nuclear envelope drives nuclear export by Exportin 1 (Xpo1), shown as an example here. Xpo1 binds directly to many substrates containing a leucine-rich nuclear export signal (NES), co-operatively with RanGTP that is abundant in the nucleus. This export complex docks to the nuclear pore complex (NPC), and is translocated to the cytoplasmic side via direct interactions of Xpo1 with FG-nucleoporins. Here, the export complex encounters two cytoplasmic regulators of the Ran GTPase cycle, Ran-binding protein 1 (RanBP1) and Ran GTPase-activating protein (RanGAP) that act together to hydrolyze GTP on Ran. As a consequence, the export complex disassembles and the export cargo is released into the cytoplasm. Since RanGTP concentrations are very low in the cytoplasm, and since Xpo1 shows very low affinity to cargoes in absence of RanGTP, Xpo1 recycles back to the nucleus on its own due to its ability to interact with FG-nucleoporins. Some export cargoes, like the pre-60S ribosomal subunit, need an adapter protein (Adapt.) that bridges its interaction with the export receptor. In the case of the pre-60S ribosomal subunit this adapter protein is designated NMD3 (Zemp and Kutay 2007), and it contains an NES that recruits Xpo1. Nuclear export proceeds as described above. The export adapter has then to be recycled to the nucleus by an import receptor. NE, nuclear envelope

(RanGAP) is restricted to the cytoplasm and these proteins concentrate at the cytoplasmic side of the nuclear envelope. In humans, the large nucleoporin RanBP2 (or NUP358, which is missing in plants) performs a similar function to RanBP1. It is localized at the cytoplasmic face of the NPC and, like RanBP1, is an example for a RanBP that is implicated in nuclear transport but has no receptor function. As a consequence of these topological arrangements, GTP hydrolysis on Ran occurs when an export complex reaches the cytoplasmic side of the nuclear envelope after its passage through the NPC. Since the affinity of exportins for their cargo substrates is very low in the absence of RanGTP, GTP hydrolysis on Ran results in the dissociation of export complexes and hence leads to the release of export cargo, RanGDP, and the exportin into the cytoplasm (Fig. 1). Complementarily, importins have very low affinity for their cargo substrates in the presence of RanGTP, and inter-

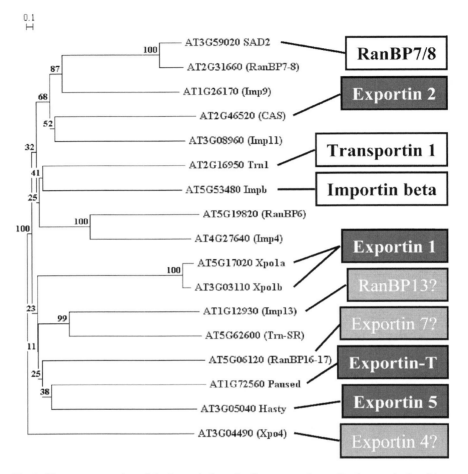

Fig. 2 How many proteins of the Importin beta family are experimentally characterized and how many of them act as exportins in plants? The phylogenetic tree shows all 17 Arabidopsis Importin beta-like nuclear transport receptors, labeled with the AGI locus designation and the Arabidopsis protein designation, if applicable, or with the designation of the human protein (*in parenthesis*) that shows highest similarity. Functionally characterized plant importins are indicated by a *boxed label*, whereas plant exportins are designated with a *dark grey* box. Putative plant exportins are given with a *question mark in a light grey box*. Functionally characterized plant nuclear transport receptors: SAD2 (Zhao et al. 2007), Exportin 2 (Haasen and Merkle 2002), Transportin 1 (Ziemienowicz et al. 2003), Importin beta (in rice; Jiang et al. 1998), Exportin 1 (Haasen et al. 1999), Exportin-T (Hunter et al. 2003), and Exportin 5 (Bollman et al. 2003; Park et al. 2005). Protein alignments were done with full-length protein sequences using ClustalW2 (www.ebi.ac.uk/Tools/clustalw2), and the phylogenetic tree was constructed with TreeCon using Poisson correction and neighbor joining, taking insertions and deletions into account (Van de Peer and De Wachter 1997). Xpo4 was used to root the tree. Distance bar and bootstrap values are given at the top left and at the internodes, respectively

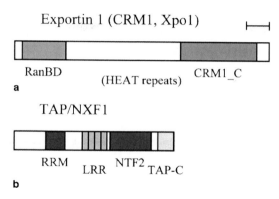

action of RanGTP with importins favors dissociation of import complexes and inhibits their stable re-formation in the nucleus. By coupling nuclear transport by Importin beta-like receptors to the irreversible hydrolysis of GTP on Ran, the characteristic features of the Ran GTPase cycle are exploited to drive directionality of these processes. In addition, the input of metabolic energy allows for accumulation of cargo substrates in one compartment against concentration gradients.

Proteins of the second class that act as nuclear export receptors are completely unrelated to RanBPs (Fig. 3). They are designated TAP (for Tip-Associated Protein)/NXF1 (for Nuclear RNA Export Factor 1) and p15/NXT1 (for NTF2-like Export Factor 1) in humans, and form TAP-p15 heterodimers to facilitate export of mRNA out of the nucleus (Cullen 2003; Rodriguez et al. 2004). These proteins are conserved in yeast, where they were designated Mex67 and Mtr2 (mRNA Transport Regulator 2), respectively. The designations TAP and p15 will be used from here on to avoid confusion. The common feature of TAP, p15 and related proteins seems to be a conserved domain, termed NTF2-like domain that is primarily important for protein–protein interaction (Suyama et al. 2000). The domain designation originates from NTF2 (Nuclear Transport Factor 2), a small protein that essentially consists of this domain, forms homodimers, and acts as a nuclear import receptor for RanGDP (Ribbeck et al. 1998; Smith et al. 1998). p15, like NTF2 is a small protein that consists of an NTF2-like domain, only. In contrast to NTF2, p15 does not form homodimers and does not interact with RanGDP. Instead, it binds to TAP, a protein containing several specific domains (Katahira et al. 2002; Wiegand et al. 2002). Apart from an NTF2-like domain that is necessary for heterodimerization with p15, TAP contains a non-canonical RNA recognition motif (RRM), leucine-rich repeats (LRRs), and the so-called TAP-C domain. This domain, as its name implies, is located at the very C terminus of the protein, belongs to the superfamily of UBA-like domains, and together with the NTF2-like domain is important for binding to FG repeat-containing nuclear pore proteins (Fribourg et al. 2001; Levesque et al. 2006). The TAP/p15 heterodimer does not make direct use of the topological information provided by the RanGTP gradient across the nuclear envelope. Thus, directionality of nuclear export of mRNA has to be achieved by other means. However, the RanGTP gradient is needed for the re-import of proteins involved in mRNA export after they dissociate from their cargo (Izaurralde et al. 1997). While there is only one protein in yeast, a small gene family encoding TAP-like proteins is present in humans (Herold et al. 2000).

2.2 Plant Export Receptors

In general, nuclear transport and the nuclear transport machinery are well-conserved between species. This is demonstrated by the findings that *Arabidopsis* and tobacco Ran, *Arabidopsis* Xpo1 (Haasen et al. 1999), Xpo-t (Hunter et al. 2003), RanGAP1 (Pay et al. 2002), and NTF2a (Zhao et al. 2006) are functional in yeast as well as by the fact that many additional plant proteins show considerable high sequence similarity to members of the vertebrate or yeast nuclear transport machinery (Merkle 2001, 2003). There are, however, also many plant-specific features, and many putative plant orthologs of the vertebrate or yeast nuclear transport machinery are not characterized functionally to date or even seem to be missing in plants.

As mentioned above, the *Arabidopsis* genome contains 17 genes that code for Importin beta-like nuclear transport receptors (Fig. 2). How many of these are exportins? *Arabidopsis* Exportin 1 is encoded by two genes that produce two very similar proteins (Xpo1a, Xpo1b) with high sequence similarity to the vertebrate and yeast nuclear export receptors CRM1/Xpo1 (Fig. 3). *Arabidopsis* Exportin 1 interacts with RanGTP and confers nuclear export of proteins that contain a leucine-rich nuclear export signal (NES; Haasen et al. 1999). The notion that the function of Exportin 1 is conserved in plants is underlined by the finding that the NES of the HIV protein Rev is recognized by *Arabidopsis* Xpo1 and that the Rev NES confers rapid nuclear export of a plant protein. In addition, both *Arabidopsis* Exportin 1 proteins contain the conserved cysteine residue that is the site of modification by leptomycin B (LMB; Haasen et al. 1999; Kudo et al. 1999). Mutation of this residue into threonine results in loss of LMB sensitivity of Exportin 1-mediated nuclear export in plant cells, like with *S. cerevisiae* Crm1p (Haasen et al. 1999 and unpublished results). Exportin 2 was the second *Arabidopsis* exportin that was functionally characterized (Haasen and Merkle 2002). Exportin 2 is also designated CAS for Cellular Apoptosis Susceptibility Protein or CSE1 for Chromosome Segregation 1 for the effects of mutations in the vertebrate and yeast genes. It is also known as Importin alpha re-exporter, which describes its immediate cellular function. Importin alpha proteins are import adapters that recruit many different karyophilic proteins containing a basic nuclear localization signal (NLS) and link them to the import receptor Importin beta for nuclear import (Hood and Silver 1998; Künzler et al.

1998; Kutay et al. 1997; Solsbacher et al. 1998). *Arabidopsis* contains at least eight genes encoding different Importin alpha proteins (Merkle 2001), and four of them (Impa1-4) were shown to interact with *Arabidopsis* CAS, which also binds to Ran (Haasen and Merkle 2002). Importin alpha and the NLS cargo are released into the nucleus after import by Importin beta, and Importin alpha has to be recycled into the cytoplasm by an export receptor since it cannot cross the NPC barrier on its own (Görlich and Kutay 1999). CAS ensures bulk recycling of Importin alpha proteins to the cytoplasm, and this function of CAS seems to be conserved in plants.

Two *Arabidopsis* exportins were identified in genetic screens as mutants showing pleiotropic developmental phenotypes, and later it turned out that the mutated genes encode Importin beta-like nuclear export receptors. Interestingly, the gene that encodes the *Arabidopsis* ortholog of the export receptor for tRNAs, Exportin-t or Xpo-t, was identified in genetic screens for mutants in different developmental pathways (Hunter et al. 2003; Li and Chen 2003), and was designated *Paused (PSD)*. *Arabidopsis* PSD/Xpo-t shares high sequence similarity with its vertebrate and yeast orthologs and was characterized for its immediate cellular function by interaction with the GTPase Ran and by complementation experiments using expression of the *Arabidopsis PSD* gene in yeast strains that carried a temperature-sensitive allele of *Los1*, the gene encoding yeast Xpo-t (Hunter et al. 2003). The second gene was also identified in a genetic screen for developmental mutants and found to encode a protein with high sequence similarity to Exportin 5 (Bollman et al. 2003). This gene was designated *Hasty (HST)* since the *Arabidopsis* mutant showed an accelerated change from the juvenile to the adult phase. However, this mutant showed also other phenotypes, and the finding that human Exportin 5 is the export receptor for double-stranded RNA (dsRNA) molecules including precursor microRNAs (pre-miRNAs) drew experiments in that direction. HST was characterized to interact with Ran, to bind to dsRNA, and to be involved in the export of miRNAs to the cytoplasm (Bollman et al. 2003; Park et al. 2005), indicating that Exportin 5 function is conserved in plants. However, there are plant-specific features that are discussed below.

Apart from these five proteins that were experimentally characterized as exportins, there are three more proteins of the *Arabidopsis* Importin beta family that are functionally uncharacterized to date and that may also act as nuclear export receptors (Fig. 2). At3g04490 shows high similarity to human Exportin 4 that functions as an export receptor for eukaryotic translation initiation factor eIF5A (Lipowsky et al. 2000). Furthermore, At5g06120 shows high similarity to human RanBP16 and RanBP17, two highly similar proteins. Since *Arabidopsis* contains only one such protein, and since RanBP16 has been designated Exportin 7 because it confines p50RhoGAP and 14-3-3s to the cytoplasm, in addition to exporting other substrates from the nucleus (Mingot et al. 2004), the *Arabidopsis* protein may also act as an exportin. For another human export receptor, Exportin 6/RanBP20 that exports profiling-actin complexes out of the nucleus (Stüven et al. 2003), there seems to be no *Arabidopsis* counterpart. Finally, there is Importin 13/RanBP13 that imports a number of different substrates into the nucleus, but was also shown to be able to act as an export receptor for eIF1A (Mingot et al. 2001). At1g12930 shows high similarity to human Importin 13. Since these speculations are solely based on similarity to human exportins functional categorization of the plant proteins has to await experimental characterization.

3   Nuclear Export of Proteins and Implication
for Plant Signaling

Not all proteins that are imported into the nucleus, however, should simply accumulate
in this compartment. To be able to perform their cellular function some proteins
have to shuttle continuously between the nucleus and the cytoplasm. Nuclear export
and import receptors of the Importin beta family are examples of such proteins, but
they are able to cross the NPC in both directions on their own due to their ability to
interact with FG-nucleoporins. All other proteins that need to shuttle between the
nucleus and the cytoplasm and that are unable to cross the NPC barrier on their own
require a nuclear export pathway to leave the nucleus after their import. Examples
of this kind of protein are the import adapters Importin alpha and Snurportin
(Görlich and Kutay 1999; Merkle 2003). Both adapters recruit import substrates in
the cytoplasm and link them to the Importin beta pathway. After nuclear import,
binding of RanGTP to Importin beta induces the dissociation of the import complexes. In contrast to the import adapter proteins, Importin beta leaves the nucleus
on its own in complex with RanGTP. Importin alpha requires the CAS nuclear
export pathway to recycle to the cytoplasm, in the case of Snurportin the export
receptor is CRM1/Xpo1 (Paraskeva et al. 1999).

CAS represents one nuclear export pathway for proteins, for which the receptor
has been characterized in plants (Haasen and Merkle 2002). To date it is not known
whether or not CAS transports any substrates other than Importin alpha proteins.
An argument for the existence of an exclusive nuclear export pathway for Importin
alpha adapter proteins is the high number of different cargo substrates including
various transcription factors and the high rate of nuclear import processes that
depend on the Importin alpha/beta heterodimer (Görlich and Kutay 1999). This
argument is in line with the importance of this nuclear import pathway for growth
and development, and it also emphasizes highly dynamic exchange of proteins and
information between the nucleus and the cytoplasm. Efficient bulk recycling of
Importin alpha proteins to the cytoplasm by CAS is thus important for effective
nuclear import of NLS proteins because this cellular process is needed at all times
during interphase. It is not known to date what kind of phenotypes might arise from
mutations in the single *Arabidopsis CAS* gene, or whether or not a null mutant
would be lethal. If vital mutants of the *CAS* gene exist in *Arabidopsis*, one would
expect a pleiotropic phenotype since an impaired functionality of CAS would result
in considerable shortage of NLS-containing proteins in the nucleus, including many
transcription factors. Many plant signaling cascades include regulated nuclear
import of a protein, for example a transcription factor, at a certain time point in
development or after an exogenous stimulus that transports information to the
genome in order to induce changes in gene expression as a response (Merkle 2003).
If this signaling molecule depends on the Importin alpha/beta heterodimer for
nuclear import, a decrease in or a failure of this response would be expected in a
plant carrying a mutant *CAS* allele. In addition, in the light of Importin alpha and
Importin beta having additional important functions during mitosis (Weis 2003),
one can envisage phenotypes related to defects in these functions as well.

Another aspect of nuclear export pathways is that they are used to ensure distinct localization of specific proteins and restrict them to the cytoplasm. This was reported for many eukaryotic translation factors (Bohnsack et al. 2002; Calado et al. 2002), RanBP1 (Görlich and Kutay 1999), and the above-mentioned proteins p50RhoGAP and 14-3-3s that are confined to the cytoplasm by Exportin 7 (Mingot et al. 2004), to mention a few examples. The finding that most human translation initiation factors, all elongation factors and the termination factor eRF1 are actively excluded from the nucleus is a strong argument against the existence of nuclear translation (Dahlberg and Lund 2004). Nuclear exclusion of these factors is accomplished by several exportins, including Exportin 1, Exportin 5, and RanBP13 (Bohnsack et al. 2002; Calado et al. 2002). In the case of RanBP1, that is also actively excluded from plant nuclei by Xpo1 (Haasen et al. 1999), this mechanism ensures that RanBP1 is able to serve as a functional marker for the cytoplasmic side of the NPC and to act, together with RanGAP, as a dissociation factor for export complexes of Importin beta family members.

Other proteins are only temporarily excluded from the nucleus or show changes in their relative abundance in the nucleus versus the cytoplasm. There are numerous examples in metazoa and in yeast that demonstrate how nucleo-cytoplasmic partitioning of specific proteins is exploited as a regulatory mechanism to control signaling to the nucleus, the classic examples are the transcription factors NF-kappaB in humans and Pho4 in yeast (Kaffman and O'Shea 1999; Turpin et al. 1999). In general, both nuclear import and nuclear export may be subject to regulation. In many cases, these proteins contain both an NLS and an NES, and their steady state localization in these two compartments is a function of the rate of nuclear import versus the rate of nuclear export. The steady state localization of such a protein may then be changed by different mechanisms that affect localization signal activity. Taking nuclear export as the example, this change may be achieved by protein modifications that interfere with exportin interaction. Alternatively, protein modifications may lead to conformational changes that mask or unmask the NES of a protein (intramolecular masking), a process that may also be induced by changes in redox potential. Finally, interaction with another protein may mask or unmask the NES (intermolecular masking) or may confine the protein to one compartment (anchoring). All these different mechanisms that result in changes in protein localization may be triggered by endogenous or exogenous signals, and hence result in regulated nucleo-cytoplasmic partitioning of specific proteins (Poon and Jans 2005).

In most reports, the nuclear transport receptors that are involved in regulated nucleo-cytoplasmic partitioning are Importin beta and Exportin 1, in yeast also the exportin Msn5 that is structurally but not functionally similar to Exportin 5. They are considered as being outstanding among nuclear transport receptors with regard to the diversity and to the high number of different cargo proteins they transport across the nuclear envelope. Exportin 1, like Importin beta, directly interacts with many different cargoes and uses adapter proteins to recruit additional cargoes or complexes. Examples of Exportin 1 adapters are NMD3 and PHAX that serve as export adapters for large ribosomal subunits and UsnRNAs, respectively (Johnson et al. 2002; Ohno et al. 2000; Thomas and Kutay 2003). *Arabidopsis* Xpo1 shares similar properties with its mammalian and yeast orthologs with regard to the high number of different transport substrates. To date, many different plant proteins were shown or are discussed to be linked directly or indirectly to Xpo1, and these proteins are implicated in different signaling pathways (Table 1). In addition, proteins encoded by plant viruses were reported to use the Exportin 1 pathway, including the nuclear shuttle protein NSP from the geminivirus Squash leaf curl virus (Carvalho et al. 2006), and RNA-3 encoded viral protein p25 (Vetter et al. 2004), and RNA-3 encoded viral protein p26 (Link et al. 2005) of Beet necrotic yellow vein virus. These findings raise the possibility that Xpo1 may be used for viral reproduction and/or spread between plant cells. The variety of different proteins in this list clearly shows that plant Xpo1 has a similar broad cargo spectrum to its orthologs in metazoa and yeast.

Recent reports of nucleo-cytoplasmic partitioning of an *Arabidopsis* signaling protein (Gampala et al. 2007; Ryu et al. 2007) and its rice ortholog (Bai et al. 2007)

Table 1 Plant cargo substrates for Exportin 1

Protein	Locus	Pathway	Reference
RanBP1	AT1G07140 AT2G30060 AT5G58590	Nucleo-cytoplasmic transport	Haasen et al. (1999)
Transcriptional repressor BZR1	AT1G75080	Brassinosteroid signal transduction	Ryu et al. (2007)
bZIP10	AT4G02640	Basal defense and cell death	Kaminaka et al. (2006)
High mobility group type B protein, HMGB	AT5G23405	Chromatin structure and remodeling	Grasser et al. (2006)
Cyclin-dependent kinase inhibitor ICK1/KRP1	AT2G23430	Cell cycle regulation	Jacoby et al. (2006)
Tomato heat shock transcription factor Hsf2a		Heat stress response	Heerklotz et al. (2001)
Rice bZIP transcription factor RSG	OS12G06520	Gibberellin signal transduction	Igarashi et al. (2001)
Tobacco and rice nucleosome assembly proteins (NAP1)		Nucleosome assembly, chromatin remodeling	Dong et al. (2005)
SR splicing factor RSZp22	AT4G31580	mRNA processing and export	Tillemans et al. (2006)
FHY1	AT2G37680	Phytochrome A signal transduction	Zeidler et al. (2004)
Chaperone DjC6	AT5G06910	Protein folding	Suo and Miernyk (2004)
Ribonuclease III-like protein RTL2	AT3G20420	Double-stranded RNA binding	Comella et al. (2007)
LHY/CCA1-like 1 (LCL1)	AT5G02840	Circadian clock	Martini et al. (2007)

serve as an excellent example to demonstrate the potential of this regulatory mechanism in plants. Brassinosteroids (BRs) are a group of phytohormones that are implicated in the regulation of plant growth and development. Plants that are defective in BR synthesis or signaling show growth characteristics in the dark that are normally associated with photomorphogenesis and remain dwarf in light as compared to wild type. BRs are perceived by a receptor kinase complex in the plasma membrane that contains BRASSINOSTEROID INSENSITIVE 1 (BRI1; Ryu et al. 2007 and references therein). Binding of BR to the receptor BRI1 results in dissociation of a repressor and dimerization of BRI1 with BRI1-ASSOCIATED RECEPTOR KINASE 1 (BAK1). This complex then activates a signal transduction pathway that modulates the activity of two related proteins that are the key transcription factors in BR signaling. Ryu et al. (2007) analyzed one of them, Brassinazole Resistant 1 (BZR1), and show that its nucleo-cytoplasmic localization and its phosphorylation status are regulated in a BR-dependent manner. BR induces rapid dephosphorylation of BZR1 in the cytosol by a plant-specific phosphatase named BSU1 and this correlates with nuclear accumulation of BZR1, where it acts as a transcriptional repressor. The nuclear import pathway of BZR1 is not clear to date, however, the presence of a basic motif in its N terminus that resembles a bipartite NLS suggests import via Importin alpha/beta heterodimers. In contrast, phosphorylation by the nuclear kinase BIN2 of several residues at two different sites of BZR1 directly inhibits DNA binding and induces nuclear export by Xpo1 and cytosolic retention of BZR1. Interestingly, 14-3-3 proteins interact with BZR1 in a phosphorylation-dependent manner and are implicated in nuclear export and/or cytosolic retention of BZR1 (Gampala et al. 2007; Ryu et al. 2007). Similar findings were reported for the bZIP transcription factor RSG in rice (Igarashi et al. 2001; Ishida et al. 2004). Taken together, BR regulates the transcriptional activity of BZR1 by controlling its nucleo-cytoplasmic partitioning by a switch involving the opposing actions of the nuclear kinase BIN2 and cytosolic phosphatases including BSU1. Such a mechanism allows for quick responses to environmental and developmental signals triggered by BR without de novo synthesis of BZR1. In addition, Ryu et al. (2007) also showed that there is a temporal and spatial regulatory component, since the kinase BIN2 shows tissue- and developmental-specific expression.

4 Nuclear Export of RNA and RNPs

4.1 Export of Small RNAs

Two genes encoding Importin beta-like nuclear transport receptors that directly bind to RNA were characterized in mutant screens in *Arabidopsis*. *Paused* (*PSD*) encodes the ortholog of the export receptor for tRNAs, Exportin-t or Xpo-t in vertebrates and *Los1* in yeast. It was identified in screens for mutations that affect meristem initiation during embryogenesis and in screens for mutations resulting in adult characteristics in the first two normally juvenile leaves (Hunter et al. 2003) as well as in a genetic modifier screen for enhancers of the weak class C loss-of-function phenotype of *hua1-1 hua2-1* double mutant *Arabidopsis* plants (Li and Chen 2003). In line with its identification in three different genetic screens, *psd* mutants show a pleiotropic phenotype. In *psd* mutants phase change is affected, the plants display adult characteristics on leaves that are juvenile in wild-type plants. They also show defects in the shoot apical meristem that were already detectable at embryo stages, and in the phyllotaxis of the inflorescence, they are delayed in the transition to reproductive development, in root growth, lateral root initiation, and they also show significantly reduced fertility (Hunter et al. 2003). Interestingly enough for a defect in an export pathway for bulk cargoes like tRNAs, *psd* plants are viable. tRNAs may be also exported from the nucleus by human Exportin 5 (Bohnsack et al. 2002; Calado et al. 2002). Plants that carry mutations in the *PSD* and in the *HST/Exportin 5* gene are also viable, show a combination of the *hst* and *psd* mutant phenotypes (see below), and are further decreased in size as compared to plants carrying a single mutation (Hunter et al. 2003). The authors speculate that this may indicate the existence of another nuclear export pathway for tRNA that is unknown to date. However, tRNAs may also leave the nucleus by passive diffusion or in a piggy-back mechanism, and these very inefficient processes may be enough for survival of single and double mutant plants. A defect in a pathway that efficiently supplies the cytosol with an important substrate like tRNA may influence all aspects of plant growth and development. The phenotypes observed in *psd* mutants may thus hint to developmental processes that are especially sensitive for an under-representation of specific protein factors.

The *HASTY* (*HST*) gene was identified in a screen for mutations that affect the transition from the juvenile to the adult phase of vegetative development as well (Bollman et al. 2003; Telfer and Poethig 1998). *hst* mutant plants also show a pleiotropic phenotype with defects in many different processes in *Arabidopsis* development, including the size of the shoot apical meristem, accelerated vegetative phase change, the transition to flowering, disruption of the phyllotaxis of the inflorescence, and reduced fertility. In addition, *hst* seedlings have an abnormally short hypocotyl and primary root. Interestingly, the leaves of *hst* plants are curled upwards, and the abaxial layer of leaf mesophyll cells resembles the adaxial layer. This defect in organ polarity was also obvious in carpels (Bollman et al. 2003), and this phenotype is reminiscent of the polarity defects in leaves caused by deregulation of miR165/miR166-control of the transcript levels of class III homeodomain-leucine zipper transcription factors (Mallory et al. 2004, and references therein). Cargo substrates for human Exportin 5 are eEF1A via tRNA (Bohnsack et al. 2002; Calado et al. 2002), small RNAs that contain a double-stranded mini-helix domain and associated RNA-binding proteins (Gwizdek et al. 2003, 2004), and precursor microRNAs (pre-miRNAs; Bohnsack et al. 2004; Lund et al. 2004; Yi et al. 2003). These findings suggested that *Arabidopsis* HST/Exportin 5 may also act in miRNA biogenesis in plants. However, there are major differences between plants and metazoa regarding the miRNA pathway. Two RNase III-like activities are involved in metazoan miRNA biogenesis that perform different functions and show different subcellular localizations (Jones-Rhoades et al. 2006). Drosha in the nucleus sets the first cut in the primary miRNA that gives rise to the pre-miRNA and already defines one end of the mature miRNA. After export of the pre-miRNA from the nucleus to the cytoplasm by Exportin 5, cytosol-localized Dicer sets the second cut that produces a duplex intermediate containing the mature miRNA and miRNA*. In *Arabidopsis*, there are four Dicer-like (DCL) proteins that are implicated in the production of small regulatory non-coding RNAs, and DCL1 is the major RNase III-like protein that is responsible for miRNA biogenesis (Jones-Rhoades et al. 2006). As a consequence, plant DCL1 is responsible for both processing steps. Since DCL1 is localized predominantly in the nucleus, plant pre-miRNAs are very short-lived intermediates and mature single-stranded miRNAs are already produced in the plant nucleus (Papp et al. 2003; Park et al. 2005). It is unclear whether HST binds to and facilitates nuclear export of single-stranded miRNAs or of the miRNA:miRNA* duplexes (Park et al. 2005). PSD does not transport miRNAs to at least partially compensate impaired HST function, since miRNA biogenesis is not affected in *psd* mutants, whereas *hst* mutants have reduced accumulation of most miRNAs. On the other hand, in *psd* mutants, the biogenesis of a tRNA was impaired, but this pathway was not affected in *hst* plants (Park et al. 2005). Does then another as yet unknown nuclear export receptor exist in *Arabidopsis* that is able to partially compensate HST malfunction? Or may miRNAs find their way to the cytoplasm by other means as well, albeit less efficiently than with a functional *HST* gene? It is possible that miRNAs exit the nucleus by association with proteins, in complex with mRNPs, or even in complex with their mRNA targets, as proposed by Park et al. (2005). It is also unknown whether or not plant miRNAs are already associated with components of the RNA-induced silencing complex in the nucleus or when they are transported to the cytoplasm. Park et al. (2005), however, found indirect evidence for nuclear activity of miRNA-dependent mRNA cleavage. Taken together, a pleiotropic phenotype is to be expected from a mutation in a nuclear export pathway that directly and indirectly affects many different cellular processes including growth and development. In addition, the *Arabidopsis DCL1* gene is subject to feedback regulation by miR162 (Xie et al. 2003). As discussed above certain *hst* phenotypes may hint at deregulated expression of specific genes and at pathways that are especially sensitive to such effects. Since so many different pathways are potentially impaired, however, conclusions are difficult.

4.2   Export of mRNA

Research from many laboratories in yeast and vertebrate cells revealed that nuclear export of mRNA is a very complex process. There is a high degree of integration of processes that are implicated in gene expression including chromatin remodeling, transcription, pre-mRNA processing, nuclear export, and onset of translation. It is thought that the tight linkage of these processes enables surveillance mechanisms to ensure that only completely and accurately processed mRNAs reach the cytosolic translation machinery and are translated into proteins. Pre-mRNA is already bound by proteins co-transcriptionally, forming large ribonucleoprotein complexes (mRNPs), including proteins that accompany the mRNA to the cytoplasm, and some of them also have important functions in mRNP export. In addition, some actively transcribed genes are positioned at the nuclear periphery where they are already tethered to the NPCs during transcription. Research on these different aspects of mRNA biogenesis in yeast and vertebrates has been covered in several excellent reviews (Cole and Scarcelli 2006; Cullen 2003; Stutz and Izaurralde 2003; Vinciguerra and Stutz 2004). The focus of this chapter is to discuss plant proteins that are implicated in mRNA export. In plants, there are several reports on genes that, when mutated, result in defects in RNA transport and/or encode proteins with high similarity to yeast or vertebrate proteins with a function in mRNA export.

During transcription, many different proteins associate with nascent pre-mRNA, among them poly(A)-binding proteins (PABPs). While yeast has one and humans three genes encoding PABPs that are highly conserved in eukaryotes, *Arabidopsis* contains a small gene family of eight members (Belostotsky 2003). Yeast Pab1 is implicated in mRNA biogenesis and export, regulation of mRNA turnover, and initiation of translation. *Arabidopsis* PAB3 binds to RNA and rescues an otherwise lethal phenotype of the yeast *pab1* mutant. In addition, PAB3 over-expression in a yeast strain carrying the *nab2-1* mutation (a gene encoding a shuttling hnRNP protein that is required for mRNA export) suppressed the export defect. This indicates that PABPs share an evolutionary conserved function in mRNA biogenesis and export (Chekanova and Belostotsky 2003). However, *Arabidopsis* PAB3 is not the ortholog of yeast Pab1 (Chekanova et al. 2001). *Arabidopsis* PABPs may recruit several other proteins to the mRNP, including other RNA-binding proteins (Bravo et al. 2005).

After capping of the 5 end of the mRNA, the cap-binding complex (CBC) forms at the cap structure and accompanies mRNPs to the cytoplasm. *Arabidopsis* contains two single genes that encode homologs of the large and the small subunit of the CBC, termed *CBC80* and *CBC20*, respectively (Kmieciak et al. 2002). *Arabidopsis ABH1/CBC80* was identified as a mutant that confers abscisic acid (ABA)-hypersensitive regulation of seed germination, stomatal closure, and cytosolic calcium increase in guard cells. The authors concluded that mRNA processing factors act as negative regulators for ABA signaling. The recent identification of the RNA-binding protein and flowering regulator FCA as one of at least two ABA receptors is in line with this (Razem et al. 2006). ABH1 is mainly localized in the nucleus at steady state, but a slight cytosolic localization was also detected, indicating nucleo-cytoplasmic shuttling in view of its putative function (Hugouvieux et al. 2002). In addition, mutations in the *ABH1* gene suppress the FRIGIDA-mediated delay in flowering (Bezerra et al. 2004). The authors show that this phenotype is caused by the inability of FRIGIDA to increase mRNA levels of the floral repressor FLC in the *abh1* mutant. Interestingly, mutations in genes encoding NUPs also show flowering phenotypes (see below). A mutation in *Arabidopsis CBC20* was also characterized phenotypically (Papp et al. 2004).

The mutant shows a pleiotropic phenotype, confers drought tolerance in plants, and, like the *abh1* mutant, shows defects in ABA signaling.

Serine/arginine-rich (SR) proteins are splicing regulators and also have a function in mRNA export. The *Arabidopsis* SR protein RSZp22 shows high intranuclear dynamics, contains an NES and shuttles between the nucleus and the cytoplasm (Tillemans et al. 2006). Moreover, proteomic analysis of the *Arabidopsis* nucleolus identified SR proteins and many components of the post-splicing exon-junction complex that functions in mRNA export and surveillance (Pendle at al. 2005). These findings suggest that plant nucleoli are implicated in these processes. *Arabidopsis* also contains four homologs of vertebrate ALY/REF proteins (Uhrig et al. 2004). They are highly conserved RNA-binding proteins that function in mRNA export as adapters to recruit other proteins to the mRNP, most notably TAP/p15 (Stutz et al. 2000; Stutz and Izaurralde 2003). Zhao et al. (2006) characterized the *Arabidopsis* NTF2 ortholog, the import receptor for RanGDP. *Arabidopsis* contains three genes encoding similar proteins, two of which interact with Ran (termed NTF2a and NTF2b), whereas the third one does not (termed NTL for NTF2-like; Zhao et al. 2006). The function of this protein is uncharacterized to date, and it is a candidate for the *Arabidopsis* homolog of vertebrate p15. In contrast, there is no obvious homolog of TAP in the *Arabidopsis* genome. A recent report by Hernandez-Pinzon et al. (2007) characterized SDE5 as a functional component in the production of *trans*-acting short interfering RNAs (tasiRNAs) that is required for transgene silencing. The sequence similarity to TAP that is discussed, however, is very weak, and in the absence of further functional data it is too early to state that SDE5 may be a putative homolog of the human mRNA export factor.

On its way to the cytoplasm, the mRNP contacts the NPC via specific NUPs. Recently, several mutations in *Arabidopsis* genes encoding NUPS have been described (reviewed in: Xu and Meier 2008). The vertebrate NUP TPR (for Translocated Promoter Region) is located at the filaments of the nuclear basket of the NPC and serves as a docking site for mRNPs. The *Arabidopsis* gene *AtTPR* was identified in a screen for suppressors of the floral repressor *Flowering Locus C* (*FLC*; Jacob et al. 2007). *Attpr* mutants are characterized by an eightfold increase of poly(A) + RNA in the nucleus. In addition, microarray analyses showed that homeostasis between nuclear and cytoplasmic RNA is disturbed, as revealed by a loss in correlation of transcript abundance, but not transcript composition, of the nuclear versus the total RNA pool. Furthermore, a pleiotropic phenotype indicates that several signaling pathways are affected, including the flowering pathway according to the design of the screen. *Attpr* mutants are early flowering, and show defects in small RNA abundance and auxin signaling (Jacob et al. 2007). Interestingly, *attpr* and *hst* mutants have similar negative effects on the abundance of many miRNAs, whereas siRNAs are not affected. This finding suggests that HST-dependent nuclear export of miRNAs needs functional AtTPR and the export complex interacts with this NUP. Xu et al. (2007) also described the characterization of a mutation in the same gene, which they named *NUA* for *NUCLAR PORE ANCHOR*. Accumulation of poly(A) + RNA and an early flowering phenotype was also described for *nua* mutants. NUA was localized to the inner surface of the nuclear envelope. Since *nua* mutants phenocopy the effects of a mutation in the gene *EARLY IN SHORT DAYS4 (ESD4)* that encodes a SUMO protease, interaction of the two proteins was tested in yeast two-hybrid experiments (Xu et al. 2007). The positive outcome of this experiment, the *nua* and *esd4* phenotypes, and reports from yeast and mammals that Mlp1/Mlp2 and TPR, respectively, also bind a SUMO protease strongly suggests that this is also the case in *Arabidopsis*, and that AtTPR/NUA and ESD4 play a role in *Arabidopsis* mRNA export.

Like in *attpr* mutant plants (Jacob et al. 2007), defects in auxin signaling were also described in the *sar3* mutant (Parry et al. 2006). This gene was characterized as *SUPPRESSOR OF AUXIN RESISTANCE 3*, and encodes another *Arabidopsis* NUP, the homolog of mammalian NUP96. Both *sar3* and *attpr* mutants are strong suppressors of the *axr1* mutant that lacks auxin sensitivity. Also similar to *attpr* mutants, *sar3* mutants are early flowering. Yet another *Arabidopsis* mutant shows accumulation of poly(A) + RNA in the nucleus. The *atnup160-1* mutant was identified in a screen for mutations that impair cold-induced transcription of a reporter gene (Dong et al. 2006). The *atnup160-1* mutation renders plants more sensitive to chilling stress. It encodes the *Arabidopsis* homolog of mammalian NUP160, and was also isolated as *sar1* mutant in a screen for suppressors of auxin resistance conferred by the *axr1* mutation (see above; Parry et al. 2006). Nuclear accumulation of poly(A) + RNA in the mutant and nuclear rim localization clearly link the *Arabidopsis* NUP160/SAR1 with the export of mRNA. Again, the flowering pathway is also affected since the *atnup160-1* mutant is early flowering.

Interestingly, in two more *Arabidopsis* mutants, export of poly(A) + RNA from the nucleus is blocked. The *cryophyte/los4-2* mutation was identified as a mutation that confers low expression of osmotically sensitive genes and, like mutations in the *Arabidopsis NUP160/SAR1* gene, shows a defect in cold signaling. It confers cold and freezing tolerance to plants, but renders them more sensitive to heat stress (Gong et al. 2005). In addition, *los4-2* mutants are hypersensitive to ABA, which is reminiscent of mutations in the *ABH1/CBC80* gene. The mutation is allelic with the *los4-1* mutation that was identified earlier (Gong et al. 2002). However, *los4-1* mutants show an opposite phenotype since this mutation renders the plants more cold-sensitive. *los4-1* mutants show accumulation of poly(A) + RNA at low and high temperatures, whereas in *los4-2* mutants RNA export is blocked at warm and high temperatures, only (Gong et al. 2005). This suggests that *los4-2* is a temperature-sensitive allele of the DEAD box RNA helicase that is encoded by this gene. The *los4* phenotypes again show that temperature and phytohormone signaling are especially sensitive to impaired mRNA export. In yeast, Dbp5 associates with mRNA early in the nucleus and accompanies it to the cytoplasmic side of the NPC where it concentrates by binding to NPC filaments. Dbp5 is thus a nucleo-cytoplasmic shuttle protein. Its ATPase activity is activated by interaction with Gle1, an RNA export factor that is also associated with the cytoplasmic side of the NPC, and inositol polyphosphate IP$_6$, and this activity leads to mRNP re-modeling and probably constitutes a crucial step for dissociation of specific RNP factors to impose directionality on mRNP export (see Fig. 4; Cole and Scarcelli 2006; Stewart 2007). Dbp5 is highly conserved between organisms, and, although *Arabidopsis* contains

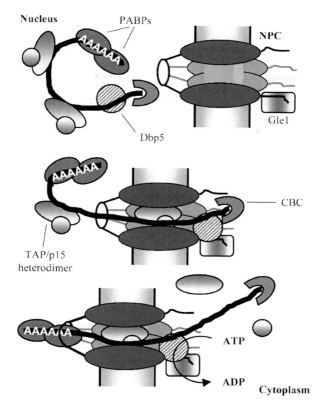

Fig. 4 Nuclear export of mRNA. Simplified model to highlight plant proteins that may be implicated in this pathway and to indicate a possible mechanism that is discussed to create directionality in mRNA export. Adapted from Cole and Scarcelli (2006) and Stewart (2007). mRNA associates in the nucleus with cap-binding complex (CBC) consisting of CBC80 and CBC20 (single genes each in *Arabidopsis*), with poly(A) binding proteins (PABPs; multigene family in *Arabidopsis* containing eight members), with Dbp5 (candidate homolog in *Arabidopsis*: LOS4), and with the RNA export receptor TAP/p15 (TAP is missing in *Arabidopsis*; the candidate homolog of p15 in *Arabidopsis* is NTL). The mRNP contacts the nuclear pore complex (NPC) via AtTPR/NUA, a nucleoporin located the nuclear basket of the NPC, and via the nucleoporin NUP160/SAR1 (not shown). Once the mRNP-associated Dbp5 has reached the cytoplasmic side of the NPC, it associates with cytoplasmic nucleoporins of the NPC and with Gle1, which in turn activates the ATP-dependent helicase of the Dbp5. This may result in the removal of RNA-binding proteins, among them the TAP/p15 heterodimer, thereby preventing the return of the transport complex to the nucleus. Many other proteins and protein complexes associate with mRNPs in the nucleus and/or in the cytoplasm but were omitted for clarity

a multigene family encoding highly related DEAD box RNA helicases, LOS4 is most similar to yeast Dbp5. In addition, Gong et al. (2005) demonstrated nuclear rim localization of LOS4-GFP.

To date, export of other RNA species is characterized in yeast and vertebrates, only. Interestingly, many RNA/RNP species including 5S rRNA, pre-40S and pre-60S ribosomal subunits, signal recognition particle (SRP; Grosshans et al. 2001), and U snRNA rely on the Exportin 1 nuclear export pathway, again highlighting the importance of this export receptor. In all of these cases, adapter proteins are implicated that recruit Exportin 1 to the respective RNP (reviewed in: Cullen 2003; Zemp and Kutay 2007). The yeast ortholog of the vertebrate RNA export receptor heterodimer TAP/p15, however, is also implicated in exporting pre-60S subunits to the cytoplasm (Yao et al. 2007). This suggests that, for important and complex cargoes, several nuclear transport receptors co-operate.

5 Conclusion

Analysis of the effect of mutations in genes encoding plant nuclear export receptors provides an effective means to characterize their function. Since many different cellular pathways are affected directly and indirectly due to impaired nucleo-cytoplasmic trafficking of different cargo substrates, however, it is difficult to draw conclusions solely based on the analysis of mutants. Not surprisingly, mutations in genes encoding nuclear transport receptors and proteins that have a general function in nuclear transport show a pleiotropic phenotype including defects in growth and development. What is more surprising is the finding that such mutations do not cause lethality. While this may not be true in general, it suggests partial redundancy of the nuclear transport machinery and/or other compensatory mechanisms that are, according to the phenotypes, of poor efficiency but allow survival under standard conditions. Having in mind their cellular function, the phenotypes of mutations in nuclear export receptors may hint to specific signaling pathways or cellular processes that are sensitive to reduced supply of (a) specific factor(s), in a temporal and spatial fashion. This may be a way to gain information on the dynamics of cellular processes. In this line of argument, it is very interesting that almost all mutations in genes encoding proteins that function in mRNA biogenesis and export identified so far show defects in the signaling of cold, stress, and phytohormones as well as in flowering.

Acknowledgements I thank all present and past members of my group, and Dorothea Haasen for critical reading of the manuscript. I apologize to colleagues whose work could not be cited directly. Support from the DFG (ME 1116/3-3, AFGN ME 1116/4-2, BIZ 7/1-2) is gratefully acknowledged.

References

Bai MY, Zhang LY, Gampala SS, Zhu SW, Song WY, Chong K, Wang ZY (2007) Functions of OsBZR1 and 14-3-3 proteins in brassinosteroid signaling in rice. Proc Natl Acad Sci USA 104:13839–13844

Belostotsky BA (2003) Unexpected complexity of poly(A) binding protein gene families in flowering plants: three conserved lineages that are at least 200 million years old and possible auto- and cross-regulation. Genetics 163:311–319

Bezerra IC, Michaels SD, Schomburg FM, Amasino RM (2004) Lesions in the mRNA cap-binding gene ABA HYPERSENSITIVE 1 suppress FRIGIDA-mediated delayed flowering in *Arabidopsis*. Plant J 40:112–119

Bohnsack MT, Regener K, Schwappach B, Saffrich R, Paraskeva E, Hartmann E, Görlich D (2002) Exp5 exports eEF1A via tRNA from nuclei and synergizes with other transport pathways to confine translation to the cytoplasm. EMBO J 21:6205–6215

Bohnsack MT, Czaplinski K, Gorlich D (2004) Exportin 5 is a RanGTP-dependent dsRNA-binding protein that mediates nuclear export of pre-miRNAs. RNA 10:185–191

Bollman KM, Aukerman MJ, Park M-Y, Hunter C, Berardini TZ, Poethig RS (2003) HASTY, the *Arabidopsis* orthologs of exportin 5/MSN5, regulates phase change and morphogenesis. Development 130:1493–1504

Bravo J, Aguilar-Henonin L, Olmedo G, Guzmán P (2005) Four distinct classes of proteins as interaction partners of the PABC domain of *Arabidopsis thaliana* poly(A)-binding proteins. Mol Genet Genomics 272:651–665

Calado A, Treichel N, Müller E-C, Otto A, Kutay U (2002) Exportin-5-mediated nuclear export of eukaryotic elongation factor 1A and tRNA. EMBO J 21:6216–6224

Carvalho MF, Turgeon R, Lazarowitz SG (2006) The geminivirus nuclear shuttle protein NSP inhibits the activity of AtNSI, a vascular-expressed *Arabidopsis* acetyltransferase reglated with the sink-to-source transition. Plant Physiol 140:1317–1330

Chekanova JA, Belostotsky DA (2003) Evidence that poly(A) binding protein has an evolutionarily conserved function in facilitating mRNA biogenesis and export. RNA 9:1476–1490

Chekanova JA, Shaw RJ, Belostotsky DA (2001) Analysis of an essential requirement for the poly(A) binding protein function using cross-species complementation. Curr Biol 11:1207–1214

Cole CN, Scarcelli JJ (2006) Transport of messenger RNA from the nucleus to the cytoplasm. Curr Opin Cell Biol 18:299–306

Comella P, Pontvianne F, Lahmy S, Vignols F, Barbezier N, DeBures A, Jobet E, Brugidou E, Echeverria M, Saez-Vasquez J (2007) Characterization of a ribonuclease III-like protein required for the cleavage of the pre-rRNA in the 3 ETS in *Arabidopsis*. Nucleic Acids Res 36:1163–1175

Conti E, Izaurralde E (2001) Nucleocytoplasmic transport enters the atomic age. Curr Opin Cell Biol 13:310–319

Cullen BR (2003) Nuclear RNA export. J Cell Sci 116:587–597

Dahlberg JE, Lund E (2004) Does protein synthesis occur in the nucleus? Curr Opin Cell Biol 16:335–338

Dong A, Liu Z, Yu F, Li Z, Cao K, Shen WH (2005) Interacting proteins and differences in nuclear transport reveal specific functions for the NAP1 family proteins in plants. Plant Physiol 138:1446–1456

Dong CH, Hu X, Tang W, Zheng X, Kim YS, Lee BH, Zhu JK (2006) A putative *Arabidopsis* nucleoporin, AtNUP160, is critical for RNA export and required for plant tolerance to cold stress. Mol Cell Biol 26:9533–9543

Fribourg S, Braun IC, Izaurralde E, Conti E (2001) Structural basis for the recognition of a nucleoporin FG repeat by the NTF2-like domain of the TAP/p15 mRNA nuclear export receptor. Mol Cell 8:645–656

Gampala SS, Kim TW, He JX, Tang W, Deng Z, Bai MY, Guan S, Lalonde S, Sun Y, Gendron JM, Chen H, Shibagaki N, Ferl RJ, Ehrhardt D, Chong K, Burlingame AL, Wang ZY (2007) An essential role for 14-3-3 proteins in brassinosteroid signal transduction in *Arabidopsis*. Dev Cell 13:177–89

Gong Z, Lee H, Xiong L, Jagendorf A, Stevenson B, Zhu JK (2002) RNA helicase-like protein as an early regulator of transcription factors for plant chilling and freezing tolerance. Proc Natl Acad Sci USA 99:11507–11512

Gong Z, Dong CH, Lee H, Zhu J, Xiong L, Gong D, Stevenson B, Zhu JK (2005) A DEAD box RNA helicase is essential for mRNA export and important for development and stress responses in *Arabidopsis*. Plant Cell 17:256–267

Görlich D, Kutay U (1999) Transport between the cell nucleus and the cytoplasm. Annu Rev Cell Biol 15:607–660

Grasser M, Lentz A, Lichota J, Merkle T, Grasser KD (2006) The *Arabidopsis* genome encodes structurally and functionally diverse HMGB-type proteins. J Mol Biol 358:654–664

Grosshans H, Deinert K, Hurt E, Simos G (2001) Biogenesis of the signal recognition particle (SRP) involves import of SRP proteins into the nucleus, assembly with the SRP-RNA, and Xpo1p-mediated export. J Cell Biol 153:745–762

Gwizdek C, Ossareh-Nazari B, Brownawell AM, Doglio A, Bertrand E, Macara IG, Dargemont C (2003) Exportin-5 mediates nuclear export of minihelix-containing RNAs. J Biol Chem 278:5505–5508

Gwizdek C, Ossareh-Nazari B, Brownawell AM, Evers S, Macara IG, Dargemont C (2004) Minihelix-containing RNAs mediate exportin-5-dependent nuclear export of the double-stranded RNA-binding protein ILF3. J Biol Chem 279:884–891

Haasen D, Köhler C, Neuhaus G, Merkle T (1999) Nuclear export of proteins in plants: AtXPO1 is the export receptor for leucine-rich nuclear export signals in *Arabidopsis thaliana*. Plant J 20:695–705

Haasen D, Merkle T (2002) Characterization of an *Arabidopsis* homologue of the nuclear export receptor CAS by its interaction with Importin alpha. Plant Biol 4:432–439

Heerklotz D, Döring P, Bonzelius F, Winkelhaus S, Nover L (2001) The balance of nuclear import and export determines the intracellular distribution and function of tomato heat stress trasncription factor HsfA2. Mol Cell Biol 21:1759–1768

Hernandez-Pinzon I, Yelina NE, Schwach F, Studholme DJ, Baulcombe D, Dalmay T (2007) SDE5, the putative homologue of a human mRNA export factor, is required for transgene silencing and accumulation of trans-acting endogenous siRNA. Plant J 50:140–148

Herold A, Suyama M, Rodrigues JP, Braun IC, Kutay U, Carmo-Fonseca M, Bork P, Izaurralde E (2000) TAP (NXF1) belongs to a multigene family of putative RNA export factors with a conserved modular architecture. Mol Cell Biol 20:8996–9008

Hood JK, Silver PA (1998) Cse1p is required for export of Srp1p/importin-a from the nucleus in *Saccharomyces cerevisiae*. J Biol Chem 273:35142–35146

Hugouvieux V, Murata Y, Young JJ, Kwak JM, Mackesy DZ, Schroeder JI (2002) Localization, ion channel regulation, and genetic interactions during abscisic acid signaling of the nuclear mRNA cap-binding protein, ABH1. Plant Physiol 130:1276–1287

Hunter CA, Aukerman MJ, Sun H, Fokina M, Poethig RS (2003) PAUSED encodes the *Arabidopsis* Exportin-t orthologs. Plant Physiol 132:2135–2143

Igarashi D, Ishida S, Fukazawa J, Takahashi Y (2001) 14-3-3 proteins regulate intracellular localization of the bZIP transcriptional activator RSG. Plant Cell 13:2483–2497

Ishida S, Fukazawa J, Yuasa T, Takahashi Y (2004) Involvement of 14-3-3 signaling protein binding in the functional regulation of the transcriptional activator REPRESSION OF SHOOT GROWTH by gibberellins. Plant Cell 16:2641–2651

Izaurralde E, Kutay U, von Kobbe C, Mattaj IW, Görlich D (1997) The asymmetric distribution of the constituents of the Ran system is essential for transport into and out of the nucleus. EMBO J 16:6535–6547

Jacob Y, Mongkolsiriwatana C, Veley KM, Kim SY, Michaels SD (2007) The nuclear pore protein AtTPR is required for RNA homeostasis, flowering time, and auxin signaling. Plant Physiol. 144:1383–1890

Jacoby MJ, Weinl C, Pusch S, Kuijt SJH, Merkle T, Dissmeyer N, Schnittger A (2006) Analysis of the subcellular localization, function, and proteolytic control of the *Arabidopsis* cyclin-dependent kinase inhibitor ICK1/KRP1. Plant Physiol 141:1293–1305

Jiang CJ, Imamoto N, Matsuki R, Yoneda Y, Yamamoto N (1998) In vitro characterization of rice importin beta 1: molecular interaction with nuclear transport factors and mediation of nuclear protein import. FEBS Lett 437:127–130

Johnson AW, Lund E, Dahlberg J (2002) Nuclear export of ribosomal subunits. Trends Biochem Sci 27:580–585

Jones-Rhoades MW, Bartel DP, Bartel B (2006) MicroRNAs and their regulatory roles in plants. Annu Rev Plant Biol 57:19–53

Kaffman A, O'Shea EK (1999) Regulation of nuclear localisation: a key to a door. Annu Rev Cell Dev Biol 15:291–339

Kaminaka H, Näke C, Epple P, Dittgen J, Schütze K, Chaban C, Holt III BF, Merkle T, Schäfer E, Harter K, Dangl JL (2006) bZIP10-LSD1 antagonism modulates basal defense and cell death in *Arabidopsis* following infection. EMBO J 25:4400–4411

Katahira J, Straesser K, Saiwaki T, Yoneda Y, Hurt E (2002) Complex formation between TAP and p15 affects binding to FG-repeat nucleoporins and nucleocytoplasmic shuttling. J Biol Chem 277:9242–9246

Kmieciak M, Simpson CG, Lewandowska D, Brown JW, Jarmolowski A (2002) Cloning and characterization of two subunits of *Arabidopsis thaliana* nuclear cap-binding complex. Gene 283:171–183

Kudo N, Matsumori N, Taoka H, Fujiwara D, Schreiner EP, Wolff B, Yoshida M, Horinouchi S (1999) Leptomycin B inactivates CRM1/Exportin 1 by covalent modification at a cysteine residue in the conserved central region. Proc Natl Acad Sci USA 96:9112–9117

Künzler M, Hurt EC (1998) Cse1p functions as the nuclear export receptor for importin a in yeast. FEBS Lett 433:185–190

Kutay U, Bischoff FR, Kostka S, Kraft R, Görlich D (1997) Export of importin a from the nucleus is mediated by a specific nuclear transport factor. Cell 90:1061–1071

Levesque L, Bor Y-C, Matzat LH, Jin L, Berberoglu S, Rekosh D, Hammarskjöld M-L, Paschal MB (2006) Mutations in TAP uncouple RNA export activity from translocation through the nuclear pore complex. Mol Biol Cell 17:931–943

Li J, Chen X (2003) PAUSED, a putative Exportin-t, acts pleiotropically in *Arabidopsis* development but is dispensable for viability. Plant Physiol 132:1913–1924

Link D, Schmidlin L, Schirmer A, Klein E, Erhardt M, Geldreich A, Lemaire O, Gilmer D (2005) Functional characterization of the beet necrotic yellow vein virus RNA-5-encoded p26 protein: evidence for structural pathogenicity determinants. J Gen Virol 86:2115–2125

Lipowsky G, Bischoff FR, Schwarzmaier P, Kraft R, Kostka S, Hartmann E, Kutay U, Görlich D (2000) Exportin 4: a mediator of a novel nuclear export pathway in higher eukaryotes. EMBO J 19:4362–4371

Lund E, Güttinger S, Calado A, Dahlberg JE, Kutay U (2004) Nuclear export of microRNA precursors. Science 303:95–98

Mallory AC, Reinhart BJ, Jones-Rhoades MW, Tang G, Zamore PD, Barton MK, Bartel DP (2004) MicroRNA control of *PHABULOSA* in leaf development: importance of pairing to the microRNA 5 region. EMBO J 23:3356–3364

Martini J, Schmied K, Palmisano R, Toensing K, Anselmetti D, Merkle T (2007) Multifocal two-photon laser scanning microscopy combined with photo-activatable GFP for in vivo monitoring intracellular protein dynamics in real time. J Struct Biol 158:401–409

Merkle T (2001) Nuclear import and export of proteins in plants: a tool for the regulation of signalling. Planta 213:499–517

Merkle T (2003) Nucleo-cytoplasmic partitioning in plants: implications for the regulation of environmental and developmental signaling. Curr Genet 44:231–260

Mingot J-M, Kostka S, Kraft R, Hartmann E, Görlich D (2001) Importin 13: a novel mediator of nuclear import and nuclear export. EMBO J 20:3685–3694

Mingot J-M, Bohnsack MT, Jäckle U, Görlich D (2004) Exportin 7 defines a novel general nuclear export pathway. EMBO J 23:3227–3236

Ohno M, Segref A, Bachi A, Wilm M, Mattaj IW (2000) PHAX, a mediator of U snRNA nuclear export whose activity is regulated by phosphorylation. Cell 101:187–198

Papp I, Mette MF, Aufsatz W, Daxinger L, Schauer SE, Ray A, van der Winden J, Matzke M, Matzke AJM (2003) Evidence for nuclear processing of plant micro RNA and short interfering RNA precursors. Plant Phys 132:1382–1390

Papp I, Mur LA, Dalmadi A, Dulai S, Koncz C (2004) A mutation in the cap binding protein 20 gene confers drought tolerance to *Arabidopsis*. Plant Mol Biol. 5(5):679–686

Paraskeva E, Izaurralde E, Bischoff FR, Huber J, Kutay U, Hartmann E, Lührmann R, Görlich D (1999) CRM1-mediated recycling of snurportin 1 to the cytoplasm. J Cell Biol 145:255–264

Park M-Y, Wu G, Gonzalez-Sulser A, Vaucheret H, Poethig RS (2005) Nuclear processing and export of microRNAs in *Arabidopsis*. Proc Natl Acad Sci USA 102:3691–3696

Parry G, Ward S, Cernac A, Dharmasiri S, Estelle M (2006) The *Arabidopsis* SUPPRESSOR OF AUXIN RESISTANCE proteins are nucleoporins with an important role in hormone signaling and development. Plant Cell 18:1590–1603

Pay A, Resch K, Frohnmeyer H, Fejes E, Nagy F, Nick P (2002) Plant RanGAPs are localised at the nuclear envelope in interphase and associated with microtubules in mitotic cells. Plant J 30:699–709

Pemberton LF, Paschal BM (2005) Mechanisms of receptor-mediated nuclear import and nuclear export. Traffic 6:187–198

Pendle AF, Clark GP, Boon R, Lewandowska D, Lam YW, Andersen J, Mann M, Lamond AI, Brown JWS, Shaw PJ (2005) Poteomic analysis of the *Arabidopsis* nucleolus suggests novel nucleolar functions. Mol Biol Cell 16:260–269

Poon IKH, Jans DA (2005) Regulation of nuclear transport: central role in development and transformation? Traffic 6:173–186

Razem FA, El-Kereamy A, Abrams SR, Hill RD (2006) The RNA-binding protein FCA is an abscisic acid receptor. Nature 439:290–294

Ribbeck K, Lipowsky G, Kent HM, Stewart M, Görlich D (1998) NTF2 mediates nuclear import of Ran. EMBO J 17:6587–6598

Rodriguez MS, Dargemont C, Stutz F (2004) Nuclear export of RNA. Biol Cell 96:639–655

Ryu H, Kim K, Cho H, Park J, Choe S, Hwang I (2007) Nucleocytoplasmic shuttling of BZR1 mediated by phosphorylation is essential in *Arabidopsis* brassinosteroid signaling. Plant Cell 19:2749–2762

Smith A, Brownawell A, Macara IG (1998) Nuclear import of Ran is mediated by the transport factor NTF2. Curr Biol 8:1403–1406

Solsbacher J, Maurer P, Bischoff FR, Schlenstedt G (1998) Cse1p is involved in export of yeast importin a from the nucleus. Mol Cell Biol 18:6805–6815

Stewart M (2007) Ratcheting mRNA out of the nucleus. *Mol Cell* 25:327–330

Stutz F, Izaurralde E (2003) The interplay of nuclear mRNP assembly, mRNA surveillance and export. Trends Cell Biol 13:319–327

Stutz F, Bachi A, Doerks T, Braun IC, Seraphin B, Wilm M, Bork P, Izaurralde E (2000) REF, an evolutionary conserved family of hnRNP-like proteins, interacts with TAP/Mex67p and participates in mRNA export. RNA 6:638–650

Stüven T, Hartmann E, Görlich D (2003) Exportin 6: a novel nuclear export receptor that is specific for profilin-actin complexes. EMBO J 22:5928–5940

Suo Y, Miernyk JA (2004) Regulation of nucleocytoplasmic localization of the AtDjC6 chaperone protein. Protoplasma 224:79–89

Suyama M, Doerks T, Braun IC, Sattler M, Izaurralde E, Bork P (2000) Prediction of structural domains of TAP reveals details of its interaction with p15 and nucleoporins. *EMBO Rep* 1:53–58

Telfer A, Poethig RS (1998) *HASTY*: a gene that regulates the timing of shoot maturation in *Arabidopsis thaliana*. Development 125:1889–1898

Thomas F, Kutay U (2003) Biogenesis and nuclear export of ribosomal subunits in higher eukaryotes depend on the CRM1 export pathway. J Cell Sci 116:2409–2419

Tillemans V, Leponce I, Rausin G, Dispa L, Motte P (2006) Insights into nuclear organization in plants as revealed by the dynamic distribution of *Arabidopsis* SR splicing factors. Plant Cell 18:3218–3234

Turpin P, Ossareh-Nazari B, Dargemont C (1999) Nuclear transport and transcriptional regulation. FEBS Lett 452:82–86

Uhrig JF, Canto T, Marshall D, MacFarlane SA (2004) Relocalization of nuclear ALY proteins to the cytoplasm by the tomato Bushy stunt virus p19 pathogenicity protein. Plant Phys 135:2411–2423

Van de Peer Y, De Wachter R (1997) Construction of evolutionary distance trees with TREECON for Windows: accounting for variation in nucleotide substitution rate among sites. Comput Appl Biosci 13:227–230

Vetter G, Hily J-M, Klein E, Schmidlin L, Haas M, Merkle T, Gilmer D (2004) Nucleo-cytoplasmic shuttling of the beet necrotic yellow vein virus RNA-3-encoded p25 protein. J Gen Virol 85:2459–2469

Vinciguerra P, Stutz F (2004) mRNA export: an assembly line from genes to nuclear pores. Curr Opin Cell Biol 16:285–292

Weis K (2003) Regulating access to the genome: nucleocytoplasmic transport throughout the cell cycle. Cell 112:441–451

Wiegand HL, Coburn GA, Zeng Y, Kang Y, Bogerd HP, Cullen BR (2002) Formation of TAP/NXT1 heterodimers activates TAP-dependent nuclear mRNA export by enhancing recruitment to nuclear pore complexes. Mol Cell Biol 22:245–256

Xie Z, Kasschau KD, Carrington JC (2003) Negative feedback regulation of *Dicer-Like1* in *Arabidopsis* by microRNA-guided mRNA degradation. Curr Biol 13:784–789

Xu XM, Meier I (2008) The nuclear pore comes to the fore. Trends Plant Sci 13:20–27

Xu XM, Rose A, Muthuswamy S, Jeong SY, Venkatakrishnan S, Zhao Q, Meier I (2007) NUCLEAR PORE ANCHOR, the *Arabidopsis* homolog of Tpr/Mlp1/Mlp2/megator, is involved in mRNA export and SUMO homeostasis and affects diverse aspects of plant development. Plant Cell 19:1537–1548

Yao W, Roser D, Köhler A, Bradatsch B, Baßler J, Hurt E (2007) Nuclear export of ribosomal 60S subunits by the general mRNA export receptor Mex67-Mtr2. Mol Cell 26:51–62

Yi R, Qin Y, Macara IG, Cullen BR (2003) Exportin-5 mediates the nuclear export of pre-micro-RNAs and short hairpin RNAs. Genes Dev 17:3011–3016

Zeidler M, Zhou Q, Sarda X, Yau CP, Chua NH (2004) The nuclear localization signal and the C-terminal region of FHY1 are required for transmission of phytochrome A signals. Plant J 40:355–365

Zemp I, Kutay U (2007) Nuclear export and cytoplasmic maturation of ribosomal subunits. FEBS Lett 581:2783–2793

Zhao Q, Leung S, Corbett AH, Meier I (2006) Identification and characterization of the *Arabidopsis* orthologs of Nuclear Transport Factor 2, the nuclear import factor of Ran. Plant Phys 140:869–878

Zhao J, Zhang W, Zhao Y, Gong X, Guo L, Zhu G, Wang X, Gong Z, Schumaker KS, Guo Y (2007) SAD2, an importin-like protein, is required for UV-B response in *Arabidopsis* by mediating MYB4 nuclear trafficking. Plant Cell 19:3805–3818

Ziemienowicz A, Haasen D, Staiger D, Merkle T (2003) *Arabidopsis* Transportin is the nuclear import receptor for the circadian clock-regulated RNA-binding protein AtGRP7. Plant Mol Biol 53:201–212

The Nucleoskeleton

Susana Moreno Díaz de la Espina

Abstract The nucleoskeleton (NSK) is the dynamic nuclear network that provides support for nuclear organization and functioning. It is composed of numerous interacting structural proteins that provide a dynamic platform for the multimeric nuclear complexes involved in the regulation of replication, transcription/splicing, chromatin remodelling, nuclear transport, signalling, formation of structural elements, etc. Its protein composition is complex. Its main components are the lamins and the lamin-associated proteins (LAPs) that form a multimeric complex or lamina; the long expandable proteins of the endoskeleton such as EAST, megator, skeletor, chromator and nuance; and actin and its associated proteins; while the nucleolar domain has its own organization and composition. This chapter focuses on the organization, composition and roles of the nucleoskeleton and summarizes the data on the organization and specific protein composition of the plant nucleoskeleton.

1 The Nucleoskeleton: An Intranuclear Frame for Chromatin and Nuclear Organization and Functioning?

In eukaryotic cells, gene organization and expression are controlled at several levels from the molecular to the higher-order organization of the genome. The nucleus is organized in structural and functional domains that have their own organization and protein composition, and perform specific functions (Albiez et al. 2006). The nuclear envelope (NE) separates the nucleus from the cytoplasm and contains the nuclear pore complexes (NPCs) that regulate bi-directional molecular transport between them. In animals, the NPCs are anchored to a complex structural lamina containing lamin-dependent protein complexes formed with different architectural,

S. Moreno Díaz de la Espina
Laboratorio de Matriz Nuclear, Departamento Biología de Plantas, Centro Investigaciones
Biológicas, CSIC, Ramiro de Maeztu 9, 28040 Madrid, Spain
e-mail: smoreno@cib.csic.es

Plant Cell Monogr, doi:10.1007/7089_2008_26
© Springer-Verlag Berlin Heidelberg 2008

regulatory and signalling partners (Gruenbaum et al. 2005; Vlcek and Foisner 2007). The lamina is essential for proper nuclear organization and functioning and has essential roles in nuclear positioning, NPC organization, intranuclear topogenesis, NE breakdown and reassembly during mitosis, chromatin organization, DNA replication, gene expression and signalling (Bridger et al. 2007; Gruenbaum et al. 2005). The NE also links the nuclear interior with the cytoskeleton (CSK) through giant protein complexes of lamins with integral proteins of the inner nuclear membrane (INM) that form protein bridges across the NE which bind cytoplasmic actin and microtubules and serve as mechanical adaptors and nuclear receptors (Houben et al. 2006; Tzur et al. 2006). Lamins also form stable complexes with internal nuclear proteins forming a highly dynamic intranuclear network. Some are architectural proteins like NuMA, titin and nesprins as well as regulatory elements such as pRb and BAF, LAPs or actin (Barboro et al. 2003; Bridger et al. 2007; Margalit et al. 2007; Vlcek and Foisner 2007).

The nuclear genome partitions into separate domains called chromosome territories (CTs) (Albiez et al. 2006; Cremer et al. 2006). Individual chromosomes occupy relatively static non-random discrete territories through interphase, with individual chromatin domains extending far from their edges into the interchromatin domains (ICDs). Nuclear organization is a major epigenetic regulator of chromatin function (Shaklai et al. 2007). The fixed relative positions of genes in the nucleus condition their spatial and temporal sequence of expression and replication. Some gene clusters associate in nuclear domains, as do the tandem repeats of rDNA genes from different chromosomes that coalesce in a single nucleolar domain (Chakalova et al. 2005; Pliss et al. 2005). Nuclear positioning of individual genes depends on transcriptional activity, with distant genes co-localizing to the same transcription factory, which supports that active genes do not de novo assemble their own transcription factories but rather migrate to pre-assembled transcription sites shared with other genes that might assemble on the underlying NSK (Chakalova et al. 2005).

The interchromatin domains (ICDs) are dynamic distinct nuclear compartments that contain the transcription, replication and repair factories and are functionally linked with the CTs (Albiez et al. 2006). They contain nuclear filaments, speckles of interchromatin granules, splicing/transcription complexes and different types of nuclear bodies with specific functions, such as Cajal bodies (CBs). The border zone between the ICDs and the compact CTs constitutes the perichromatin domain that contains decondensed chromatin engaged in replication, transcription and co-transcriptional splicing (Fakan 2004).

The nucleolus, the most conspicuous nuclear domain, houses the rDNA genes, uses specific transcriptional and post-transcriptional machineries, has its own organization and performs specific functions (Shaw and Doonan 2005). Besides rDNA expression, the nucleolus is also essential for snRNP maturation, nuclear trafficking, cell-cycle regulation, and sequestration of proteins (Shaw and Doonan 2005). The nucleolus has specific subdomains for rDNA transcription and replication and later assembly of preribosomal particles, and other less well-defined elements, such as fibrillar centres, associated bodies and cavities (Kruger et al. 2007; Pliss et al. 2005; Shaw and Doonan 2005). The architectural organization of the nucleolus is epigenetically regulated (Espada et al. 2007).

Nuclear organization depends on protein modification and affects gene expression (Espada et al. 2007; Heun 2007). The chromatin is organized in loop domains, attached to the NSK at their bases by specific DNA regions called matrix attachment regions (MARs), which play crucial functions in chromatin organization, gene expression and transgene stability. Heterochromatin, a domain especially prominent in plants, forms large clusters to which temporarily silenced genes associate (Albiez et al. 2006). Nuclear functioning relies on a constant flow of proteins and RNAs between compartments, and interactions between specific addressing protein domains are essential for intranuclear topogenesis. The complexity of genome and nuclear organization suggests the existence of an underlying dynamic structural framework that would contribute to spatially coordinate the multiple nuclear components and factors, and to establish its general organization. Micromanipulation demonstrates that subnuclear structures and chromatin are interconnected. Besides, after NSK isolation protocols the distinct nuclear domains remain in place after removal of soluble proteins and chromatin (Nickerson 2001).

The discovery of expandable proteins in metazoa involved in the assembly of a contractile endoskeleton and the direct observation of their incorporation into this structure in living cells by dynamic fluorescence microscopy have resolved the controversy about the underlying NSK (Nalepa and Harper 2004; Nickerson 2001). Nevertheless its exact nature, composition and roles are not well defined, although the NSK is thought to mediate nuclear structure, DNA replication and transcription, binding of splicing factors, chromatin folding, nuclear envelope anchoring, etc.

NSK, nuclear matrix and nuclear scaffold are the terms used to refer to this intranuclear non-chromatin network that is observed in fixed cells, extracted nuclear structures and also in vivo (Barboro et al. 2003; Moreno Diaz de la Espìna 1995; Nickerson 2001), whose molecular identity remains largely unknown. The term NSK will be used to refer to this structure from here onwards.

1.1   Ultrastructural Organization and Composition

The NSK forms a branched filamentous network with a consistent architecture from mammals to plants. It contains a peripheral lamina, an endoskeleton formed by a network of intermediate filament (IF)-like core filaments of unknown composition, nucleolar- and nuclear body-remnants and interchromatin granule clusters (Fig. 1a; Barboro et al. 2003; Nickerson 2001; Yu and Moreno Díaz de la Espina 1999). Resinless electron microscopy (EM) of cells after either electroelution of chromatin, or sequential extraction with detergent, DNase and 2M salt, revealed a branched internal NSK connected to the CSK. It is made of 11-nm filaments with an axial repeat of 23 nm similar to IFs (Jackson and Cook 1988; Nickerson 2001), forming an RNA-stabilized intranuclear web (He et al. 1990) that consists of 3–5-nm lamin filaments to which NuMA islands are anchored (Barboro et al. 2003). Recently, a cell-type specific mRNA-like component of the NSK, Gomasu, has been reported (Sone et al. 2007).

This filamentous network is composed of many interacting structural proteins that provide dynamic platforms for the protein–protein and nucleic acids–protein

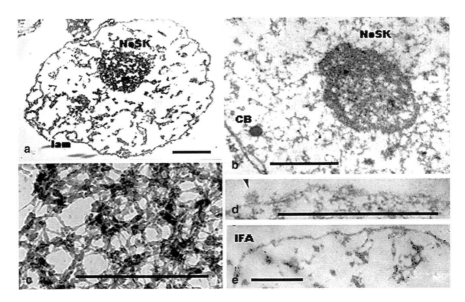

Fig. 1 Ultrastructure of the nucleoskeleton of onion meristematic root cells. **a**: Conventional EM section of an isolated NSK displaying its overall organization with a peripheral lamina-like structure (lam), the endoskeleton network and the nucleolar skeleton (NoSK). **b**: Higher magnification of an NSK showing the delicate organization of the endoskeleton, NoSK and a residual Cajal body (CB). **c**: Resinless section displaying the intricate network of the beaded filaments of the endoskeleton. **d**:H igher magnification of the lamina displaying an associated pore complex (*arrow*). **e**:T he universal anti-IF antibody IFA localizes IF-type proteins in the lamina and the endoskeleton. *Bars* = 1 μm

interactions involved in genome organization, DNA replication and repair, chromatin remodelling, transcription and co-transcriptional splicing, signalling, transport, etc., and houses the corresponding macromolecular complexes (Bettinger et al. 2003; Shumaker et al. 2003; Vlcek and Foisner 2007). The proteomic analysis reveals a complex composition of the isolated NSK whose profile varies according to the cell type, differentiation state, extraction procedure, etc. Its main structural protein components are members of the same families as those in the CSK: lamins, actin, and their associated proteins (Gruenbaum et al. 2005; Shumaker et al. 2003).

1.1.1 The Lamina

The lamina is a network of fibrous polypeptides of lamins and a complex set of lamin-binding proteins forming the nuclear exoskeleton. Lamins are its major components that also form stable complexes in the nuclear interior. They are type V intermediate filament proteins with a long α-helical coiled-coil rod domain that mediates head-to-tail dimerization and higher-order assembly; a short N-terminal head domain and a globular tail domain containing the nuclear localization signal

(NLS), an Ig fold and a terminal CAAX sequence, which is a target for isoprenylation and carboxyl methylation (Gruenbaum et al. 2005; Vlcek and Foisner 2007).

Unicellular organisms lack lamin genes, as do plants. Invertebrates have single B-type lamin genes; except for *Drosophila* that also have an A-type one (Melcer et al. 2007). Vertebrates have three genes for lamins that encode for seven proteins, belonging to types A and B that have different sequences, biochemical features and expression patterns, and do not form heteropolymers. A-type lamins A and C are generated by alternative splicing of a single gene, have neutral isoelectric points, are soluble during mitosis and incorporate to the reassembling lamina later than B-type lamins. B-type lamins B1 and B2 are essential, encoded by different genes, have acidic isoelectric points and remain membrane-bound during mitosis. Both types are translated as pre-lamins and undergo modifications of the C-terminal CAAX box including farnesylation, cleavage of the AAX tripeptide and carboxymethylation. A and B lamins have separate assembly pathways in vivo, but their specific architectural organization is unknown.

Lamins bind to DNA and its associated proteins but also interact with many nuclear membrane and nucleoplasmic LAPs, forming polymers that constitute the lamina and confer mechanical strength to the nucleus (Table 1; Bridger et al. 2007; Dorner et al. 2007; Houben et al. 2006; Schirmer and Foisner 2007; Vlcek and Foisner 2007).

Some of the LAPs are architectural partners that form stable complexes in interphase and have known structural roles. Amongst them are integral membrane proteins such as the LEM-domain proteins LAP2, lamin B receptor (LBR), emerin, LEM2 and MAN1; the SUN-domain proteins; titin; actin that interacts with emerin and spectrin and proteins of the nesprin family with a modular structure containing multiple spectrin repeats and calponin homology (CH) domains for actin binding, that enable them to interact with different components. The small isoforms of nesprin-2 bind to lamin A and emerin in the lamina. Others are chromatin partners such as histones and BAF; gene regulators such as several transcription factors and repressors like MOK2 and pRb; signalling proteins such as protein kinase C (Dorner et al. 2007; Schirmer and Foisner 2007; Vlcek and Foisner 2007). LAPs function as assembly elements for multiprotein complexes integrating structural and regulatory proteins, splicing factors, epigenetic modifiers and signalling molecules, indicating that the lamin-LAP complexes may serve as scaffold structures for nuclear architecture and nuclear reassembly after mitosis; large-scale chromatin organization, gene regulation and signalling (Dorner et al. 2007; Margalit et al. 2007; Schirmer and Foisner 2007).

The lamina also links the nuclear interior with the CSK through complexes of lamins with integral proteins of the INM such as LAPs, emerin and LBR; transmembrane proteins of the INM like SUN-domain proteins; and large nesprin isoforms of the outer nuclear membrane (ONM) containing the KASH-domain that interact with all the three major cytoskeletal components: actin, intermediate filaments and microtubules. The linking of KASH-domain proteins to mechanical internal nuclear proteins occurs through A-type lamins and emerin. Nuclear actin and spectrin filaments crosslinked by protein 4.1 form a cortical network anchored by emerin (Holaska et al. 2004; Tzur

Table 1 Main components of the NSK in plants and animals

Nucleoskeleton domain	Animals	Plants
Lamina	Lamins A–C	Lamin-like proteins
	Lamin B	NIFs
	Emerin	NMCP1
	LBR	NMP1
	LEM-2	Spectrin-like proteins
	MAN-1	
	LAP1	
	LAP2β	
	Nesprin 1α	
	SUN domain proteins	
Endoskeleton	Lamin A	Lamin-like proteins
	Lamin B	NIFs
	LAP2 β	NMCP1
	Nesprin2	Spectrin-like proteins
	NuMA	NuMA-like proteins
	Actin	Actin
	Titin	MAP190
	EAST	P 105
	Megator	MFP1
	Skeletor	NALP1
	Chromator	NMP1
	P 4.1	AHP1
		AHL1
		AHM1
Nucleolar skeleton	Nucleolin	Nucleolin
	Fibrillarin	Fibrillarin
	SURF-6	SURF-6
	Skeletor	Actin
	Nop 25	ASE-1
		UBF
		MARBP1
		MARBP2
		AtMARBP61
		NO-76

et al. 2006; Vlcek and Foisner 2007). All of them form a link complex between the NSK and CSK that connect cytoplasmic and nucleoplasmic activities (Houben et al. 2006; Tzur et al. 2006; 2006; Vlcek and Foisner 2007).

1.1.2 The Endoskeleton

The endoskeleton is less defined than the lamina in terms of both protein composition and function. Its more relevant components besides internal lamins and their associated LAPs, are the giant proteins with extended coiled-coil domains that form an expandable network in the ICDs, such as megator and EAST; the proteins associated to chromatin and chromosomes, such as chromator and skeletor; NuMA; actin and some nuclear actin-related proteins (ARPs) and actin-binding proteins (ABPs) (Table 1). All of them have important functions in chromatin organization and remodelling, transcription and RNA processing, and intranuclear transport; and some are also components of the macromolecular complex that forms the mitotic spindle matrix (Fabian et al. 2007; Zheng and Tsai 2006).

Lamins form intranuclear scaffolding multiprotein complexes involved in crucial nuclear functions, such as chromatin organization, DNA replication, transcription, apoptosis, etc. (Dorner et al. 2007; Vlcek and Foisner 2007). Fluorescence recovery after photobleaching reveals that they are more mobile than those in the lamina, but the mechanisms of their internal targeting or assembly are not clear. Little information exists about the structural and functional features of the internal lamins. A subfraction of A-type lamins are assembled into tight intranuclear foci and form a lamin internal RNA-dependent scaffold associated with NuMA (Barboro et al. 2002; Dorner et al. 2007), suggesting the existence of a functional interactive intranuclear lamin network.

A small subfraction of A-type lamins and LAP2α form stable complexes in intranuclear foci that also contain specific LAPs, transcription factors and regulators (Bridger et al. 2007). LAP2α is a nucleoplasmic LEM-domain protein that interacts directly with DNA and BAF, an essential protein involved in nuclear assembly, chromatin structure and gene expression (Margalit et al. 2007), and mediates LAP2α interaction with chromatin that would promote heterochromatin formation. The interaction with another LAP2α partner, LINT-25, would involve the complexes in cell-cycle regulation (Naetar et al. 2007). The lamin A/C-LAP2α complexes also interact with retinoblastoma protein (pRb) and actin. Altogether, these complexes would have potential functions in chromatin organization, gene expression and cell-cycle regulation in different pathways mediated by their different interactions (Dorner et al. 2007).

The expandable giant protein Titin with three isoforms higher than 1 Mda is a partner of nucleoplasmic A and B lamins in Hela cells. It contains multiple Ig-folds, fibronectin-type 3 domains, and one PEVK domain and is essential for mitotic chromosome condensation and segregation. Titin might link lamins to chromatin and/or nuclear actin during interphase, and its regulated detachment from lamins could help coordinating chromosome condensation and nuclear disassembly (Zastrow et al. 2006).

The giant Nesprin-2 isoform NUANCE, consisting of paired N-terminal CH domains, a central extended rod domain and a C-terminal KLS domain, co-localize with lamins at intranuclear foci and heterochromatin and binds nuclear actin polymers and could contribute to actin-dependent nuclear functions (Zhang et al. 2005).

EAST is a 253-KDa component of the expandable nuclear ICD containing seven potential NLSs and 12 potential PEST proteolytic signals, but no other previously characterized functional domains. EAST promotes the preferential accumulation of actin and CP60 in the nucleoplasm and modulates the spatial arrangement of chromosomes (Wasser and Chia 2000, 2005). EAST is in the same protein complex as Megator (Mtor), an ortholog of mammalian Tpr, a nuclear pore complex component (Qi et al. 2005). Mtor is an essential 260-KDa protein with an N-terminal long coiled-coil domain responsible for self assembly and polymerization, and a C-terminus involved in nuclear and spindle matrix targeting and localization. In interphase both Mtor and EAST interact to form an endoskeleton (Qi et al. 2004, 2005).

Skeletor is an 81-KDa essential protein with a bipartite NLS and serine- and histidine-rich regions, but without any significant homologies. It is not a structural component, but forms part of a multiprotein complex. Skeletor directly interacts with chromator, a 130-KDa essential chromodomain protein. The Chromator–Skeletor complex functions in two different molecular complexes, one associated with the spindle matrix and the other with nuclear chromatin in interphase. Chromator has a functional role in interphase chromatin organization and chromosome segregation (Rath et al. 2004, 2006; Walker et al. 2000). In interphase both proteins localize to chromosomes and chromatin, although skeletor is also present in the nucleolus.

NuMA is a long, 200–240-kDa conserved vertebrate coiled-coil protein. No NuMA orthologs are found in invertebrates (Abad et al. 2004), although a paralog with similar domain architecture has been reported in *Drosophila* (Bowman et al. 2006). NuMA has a very long central a-helical coiled-coil domain that mediates dimerization, a CH domain in the N-terminus and a globular C-terminus with the NLS; binding sites for LGN, tubulin and protein 4.1, several consensus sites for different protein kinases and multiple alternative splicing sites (Abad et al. 2007; Harborth et al. 2000). Its C-terminus controls the oligomerization of NuMA dimers into arm oligomeres by self-assembly, while the N-terminus controls the assembly of the latter into lattices probably by post-translational modifications by specific protein kinases and phosphatases, co-factors and binding proteins (Harborth et al. 1999). NuMA minilattices in the endoskeleton associate to internal lamins depending on RNA (Barboro et al. 2002), providing mechanical stability and compartmentalization.

NuMA has several interacting partners in nuclear multiprotein complexes, such as lamins; the transcription factor GAS41; protein 4.1 that stabilizes its interaction with actin, and the protein remodelling complex INI1 that contain ARPs via its CH domain (Mattagajasingh et al. 1999).

The interphasic functions of NuMA are not well defined. Its flexibility, capacity to form lattices and ability to bind to different structural and regulatory proteins and to MAR sequences, as well as its association with the core filaments of the endoskeleton, indicate that NuMA plays a structural role, defines nuclear shape and mechanical rigidity and has an active role in the reformation of the postmitotic nucleus (Barboro et al. 2003). Its interaction with GAS41 involves NuMA in the regulation of gene expression, probably providing a framework for transcription (Harborth et al. 2000). NuMA is also involved in chromatin organization, differentiation and apoptosis (Abad et al. 2007).

Some endoskeletal components integrate in the mitotic spindle matrix independently of MT polymerization, playing important roles in its assembly and function. Among them are EAST, Mtor and NuMA from the ICDs, the chromosomal proteins Skeletor and Chromator, titin and a small subfraction of lamin B (Fabian et al. 2007; Wasser and Chia 2005; Zheng and Tsai 2006). NuMA has also several interacting partners in the multiprotein complexes of the mitotic spindle, like Arp1 that mediates NuMA association to the dynein–dynactin motor protein complex; LGN, the mammalian Pins homolog that plays a role in spindle pole organization and tubulin (Du et al. 2002).

The pool of nuclear actin contains monomeric, oligomeric and polymeric forms with unconventional conformations (Jockusch et al. 2006; McDonald et al. 2006; Schonenberger et al. 2005) that might contribute to its multiple nuclear functions in transcription, chromatin remodelling, transport and signal transduction between nucleus and cytoplasm, nuclear structure, etc. (Bettinger et al. 2004). Actin plays an integral role in nuclear function and is associated with the nucleoskeleton. Many functional and structural nuclear proteins have CH-domains for actin association. The nuclear ABPs such as nuclear myosin I (NMI), profilin, cofilin, emerin, lamin A, p4.1, nesprins and exportin 6 control the dynamics of actin in the nucleus (Blessing et al. 2004).

Actin is an integral component of several nuclear macromolecular complexes. In transcription complexes actin is essential for the formation of the preinitiation complexes of the polymerases, and during elongation polymeric actin acts as a molecular motor with NMI necessary for maintenance of active transcription (Ye et al. 2008). Actin is also involved in RNA processing and associates to hnRNP proteins. Actin–profilin complexes are functional components of the spliceosome that co-localizes with Sm and SR proteins in nuclear speckles and CBs, and is also involved in the export of small nuclear ribonucleoproteins through the NPC. In chromatin remodelling and histone acetyl transferase complexes, monomeric actin functions in actin/Arp modules that interact with chromatin and other proteins. In them, the solubility of chromatin modifying proteins depends on the polymerization state of actin (Blessing et al. 2004; Chen and Shen 2007).

As a nucleoskeletal component actin associates with lamin A, emerin, protein 4.1, and nesprins. These last proteins have the ability to bind both lamin A and actin and could be bridging elements between the two types of important nucleoskeletal components in the lamina and the endoskeleton, contributing to the formation of a well-organized actin-containing structural network in the nucleus (Kiseleva et al. 2004; Shumaker et al. 2003). Nuclear actin and p 4.1 co-localize in intranuclear filaments that are necessary for proper nuclear assembly in vitro (Krauss et al. 2003). The identification of more actin- and Arp-binding nuclear proteins would provide important information about the role of actin in nuclear architecture and the interconnections of chromatin remodelling, transcription and RNA processing complexes with the nucleoskeleton.

1.1.3 The Nucleolar Skeleton

The nucleolus is a complex subnuclear domain with specific subdomains housing the machineries for rRNA transcription, processing and preribosome assembly. Its stability after isolation, and the specificity of its ultrastructural organization and protein composition in relation to other nuclear domains (Lam et al. 2004; Pendle et al. 2005; Shaw and Doonan 2005), suggest that the nucleolus is a self-assembly structure.

A differentiated residual nucleolar skeleton (NoSK) with an exclusively fibrillar structure, has been demonstrated by EM (Nickerson 2001), but very little information exists about its protein composition and molecular organization. Most of the structural proteins of the endoskeleton, except for skeletor, are not present in the NoSK. A few proteins have been reported in this structure and they mediate RNA–protein and protein–protein interactions in the nucleolus (Table 1), such as SURF-6, a family of nucleolar proteins with a distinct conserved C-terminal domain mediating these interactions (Magoulas et al. 1998; Polzikov et al. 2005) and nucleolin, a conserved multifunctional nucleolar protein involved in all steps of ribosome biogenesis (Goztmann et al. 1997).

Recently, some proteins involved in the assembly of the nucleolus have been identified. Most of them are RNA-binding proteins associated with rRNA transcription and processing machineries, such as Nop 25, whose N-terminus binds to nucleolar components to maintain nucleolar assembly (Suzuki et al. 2007) and UBF, a primary architectural element of active NORs (Prieto and McStay 2007). Two RNA binding proteins from the FRGY2/YB1 family that contain N-terminal motifs for sequence-specific RNA binding and a C-terminal BA island for non-selective RNA association, disassemble the nucleolus by sequestering the multifunctional nucleolar phosphoprotein B23 (Gonda et al. 2006). Interactions of nucleolin with other nucleolar proteins such as B23 and fibrillarin are also important for nucleolar structure (Ma et al. 2007). SURF-6 has a distinct domain that mediates protein–protein interactions, co-localizes with B23 and fibrillarin, and has high affinity for nucleic acids (Polzikov et al. 2005). Besides, two members of the condensin complex (Peg7 and XCAP-E) co-localize with B23 and fibrillarin in the nucleolar granular component suggesting a role in spatial organization of the nucleolus (Uzbekov et al. 2003).A ll these results suggest that both protein–protein and RNA–protein interactions are involved in the organization of the NoSK, and that the multifunctional phosphoprotein B23 could be an important mediator of nucleolar disorganization. Epigenetic modification by DNA methyltransferase 1 (Dnmt1) is also involved in the spatial organization of the nucleolar structure removing fibrillarin and Ki-67 protein from this structure (Espada et al. 2007).

2   The Plant Nucleoskeleton

2.1   *Organization and Ultrastructure of the Plant NSK*

Monocot and dicot plants have an underlying NSK similar to that in animal cells. It consists of a well-organized peripheral lamina-like structure with attached pore complexes; an endoskeleton formed by a branched network of knobbed filaments displaying associated interchromatin granules; a compact fibrillar NoSK and other distinct subdomains such as CB and MFP1-bodies (Fig. 1a,b; Masuda et al. 1993, 1997; Mínguez and Moreno Díaz de la Espina 1993; Moreno Díaz de la Espina 1995; Yu and Moreno Díaz de la Espina 1999).

Although the ultrastructural organization of the plant NSK is well established, information about its components is very scarce. The *Arabidopsis* genome contains genes for actin and a calponin–spectrin superfamily (Meagher and Fechheimer 2003),b uthomo logs of most animal nucleoskeletal proteins cannot be identified in plants by sequence homology searches. Nevertheless, the structural and functional similarities of the animal and plant nucleus and NSK, and the presence of immunologically related nucleoskeletal proteins in plants, suggest that plants have functional equivalents of the known animal nucleoskeletal proteins that have diverged a lot from their animal counterparts.

The NSK proteome of *Arabidopsis* is a complex with up to 300 spots from which 36 have been identified by mass spectroscopy. The fraction is enriched in proteins of the nucleolar domain such as fibrillarin, nucleolin and the MAR-binding proteins Nop 56 and Nop 57 (Calikowski et al. 2003). The use of sera against animal nucleoskeletal proteins in plant nuclei unveiled the presence of proteins immunologically related to lamins (McNulty and Saunders 1992; Mínguez and Moreno Díaz de la Espina 1993; Moreno Díaz de la Espina 1995), spectrin repeat-proteins (de Ruijter et al. 2000; Perez Munive et al., unpublished results), NuMA (Yu and Moreno Díaz de la Espina 1999), actin (Cruz et al. 2007), nucleolin (Mínguez and Moreno Díaz de la Espina 1996), etc.

2.2   The Plant Lamina

The plant NSK contains a lamina ultrastructurally similar to the vertebrate's one that contains the pore complexes and is associated with the INE and endoskeleton (Fig. 1a, d; Masuda et al. 1993, 1997; Mínguez and Moreno Díaz de la Espina 1993; Moreno Díaz de la Espina 1995). The plant lamina contains IF-type proteins and proteins antigenically related to lamins (Table 1; Fig. 1e; Blumenthal et al. 2004), although the plant genomes lack orthologs of lamins and their associated proteins. The lack of lamin orthologs doesn't mean that plants lack lamin functions, as lamins are involved in fundamental nuclear processes (see Sect. 1.1.1) that are similar in vertebrates and plants. Using A/C and B2 lamins as query sequences, several potential *Arabidopsis* proteins with moderate similarity were found. The alignment is predominantly with the α-helix rod domain, but some of the plant sequences also contain a homolog of the NLS (Meagher and Fechheimer 2003). The plant lamin-like proteins would have conserved domains necessary for essential lamin functions, but also other plant-specific ones can be functional rather than structural homologs of lamins. This occurs in coilin, the organizing protein of CBs, that in *Arabidopsis* is encoded by a distant coilin homolog and shows limited homology with the vertebrate proteins but has the same nuclear localization and functionality (Collier et al. 2006).

Monocot and dicot plants have NE-associated coiled-coil proteins sharing antigenic determinants with vertebrate lamins that also have similar Mr (molecular mass), *p*I (isoelectric point) values and nuclear distribution in both the exoskeleton and endoskeleton (Figs. 1e and 2a,e; McNulty and Saunders 1992; Mínguez and Moreno Díaz de la Espina 1993), and are specifically cleaved in apoptosis (Chen et al. 2000), but their sequences are not available.

Although lamin-associated proteins are not conserved in plants, BY-2 tobacco cells transformed with an N-terminal fragment of the LBR that anchors the protein to the INM, direct the transformed protein to the NE. Binding mutants of the protein have altered targeting and retention suggesting that the translocation mechanism for the LBR is conserved in plants that also have proteins that weakly bind to it (Graumann et al. 2007).

Besides lamin-like proteins, the plant lamina contains other plant-specific proteins that could be structural components of this structure such as nuclear intermediate filament proteins (NIFs) (Blumenthal et al. 2004), NMCP1 (Masuda et al. 1993, 1997), and NACs (nuclear acidic coiled-coil proteins) (Blumenthal et al. 2004).

At least three NIFs sharing antigenic determinants with chicken lamins, human keratins and IFs have been detected in the NSK of pea, *Arabidopsis* (Blumenthal et al. 2004) and onion (Pérez-Munive et al., unpublished results). They have Mr values of 65, 60 and 57 kDa and display multiple isoelectric forms with pIs between 4.8 and 6.0. Peptide sequencing showed short regions of similarity with lamins B and A that would explain their crossreactivity, with keratins and with *Arabidopsis* proteins containing long coiled coils such as NAC, FPP3 and MFP1 (Blumenthal

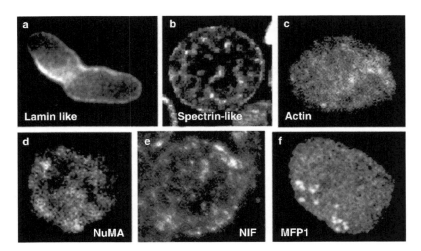

Fig. 2 Specific distribution of some protein components of the onion NSK revealed by immunofluorescent staining. **a–b**: nuclei. **c–f**: NSK. **a**: Lamin-like proteins localize to the lamina-like structure and a delicate intranuclear network. **b**: Spectrin-like proteins have a discontinuous distribution in the lamina and also in the ICDs. **c**: Actin reveals a delicate endoskeleton similar to that of NuMA (**d**), with some discrete foci. **e**: NIFs distribute in the lamina-like structure and the endoskeleton. **f**: MFP1 associates with the filaments of the endoskeleton and is enriched in specific residual nuclear bodies

et al. 2004), although the last three proteins have predicted Mrs larger than the pea NIFs. The purified NIFs form 6–12 filaments in vitro. In situ they show a peripheral and internal nuclear distribution (Fig. 2e; Blumenthal et al. 2004). Thus, plant NIFs although different from animal lamins could have lamin-like regions and be functional homologs of lamins.

NMCP1 is a 134-kD conserved protein of the carrot NE, which shows similarity to myosin, tropomyosin and IFs. It has a central coiled-coil domain, flanked by non-helical short head and long tail domains. The terminal domains contain potential NLS and many recognition motifs for protein kinases, including the CDK1 kinase and PKC. NMCP1 has a similar pI (5.6–5.8) to lamins, although its size is almost twice that of lamins (Masuda et al. 1993). Orthologs of NMCP1 are also present in the *Arabidopsis* and rice genomes but not in mammals or yeast (Moriguchi et al. 2005). The sequenced NMCP1 is a structural protein of the plant lamina, although neither the phylogenetic relationship between NMCP1 and lamins nor its polymerization abilities, its association with integral nuclear membrane proteins and ability to bind DNA are resolved. An *Arabidopsis* gene family (LINC) encoding proteins with extended coiled-coil domains related to carrot NMCP1 has been described. LINC1 localizes to the nuclear periphery, LINC2 is nucleoplasmic, LINC3 nucleolar and LINC4 plastidic. Mutation of LINC1 and LINC2 affect nuclear size and organization, chromocentre number and also whole plant morphology, and both are important determinants of plant nuclear structure (Dittmer et al. 2007).

Significantly related sequences to the members of the calponin–spectrin superfamily, were detected in the *Arabidopsis* database, but none of them are clear homologs of the query sequences (Meagher and Fechheimer 2003). Spectrin-like proteins of 220–240 and 60 kDa have been detected in the nucleus of carrot, pea (de Ruijter et al. 2000) and onion (Fig. 2b) cells by immunolabelling. Their distribution is discontinuous in the nuclear periphery. In the ICDs their distribution is either spotted or track-forming, they are practically absent from the nucleolus, and show partial co-localization with actin (Pérez-Munive et al. unpublished results). Spectrin-like proteins are also intrinsic components of the NSK (de Ruijter et al. 2000; Pérez-Munive et al., unpublished results). According to their Mr values and distribution, the spectrin-like proteins could be homologous to nesprins (see Sect. 1.1.2) that play an important structural role in the NSK.

2.3   The Endoskeleton

Plant cells have a well developed and highly stable endoskeleton that resists RNase digestion, and also LIS extraction without heat stabilization (Moreno Díaz de la Espina 1995). It basically corresponds to the residual ribonucleoprotein network of the ICDs (Fig. 1a,b). The endoskeleton is composed of a branched network of beaded filaments with 25 ± 2.5-nm knobs spaced by 15 ± 2-nm interknobs as revealed by EM of resinless sections (Fig. 1c; Yu and Moreno Díaz de la Espina

1999). In EM of conventional sections it appears as a loose fibrogranular network containing IGs that connect the lamina with the residual nucleolus and nuclear bodies (Fig. 1b,d).

In contrast with its ultrastructure, the protein composition of the plant endoskeleton is poorly defined. Its more relevant components are the internal lamin-like proteins, NIFs and spectrin-like proteins that are major components of this structure as stated above, and distribute in an endoskeletal network but not in the NoSK (Fig. 2a,b,e). In addition, large proteins that associate to both the endoskeleton and the spindle matrix during mitosis such as NuMA and MAP190, actin, and several plant specific proteins like DNA- and MAR-binding proteins, ankyrin3, and coiled-coil proteins have been reported in the plant endoskeleton (Table 1). No expansible proteins similar to those in the animal endoskeleton have been reported in plants, although recently NUA, the plant homolog of mammalian and *Drosophila* Tpr or Mtor (see Sect. 1.1.2) has been reported in *Arabidopsis* with a similar Mr. It is nuclear in interphase and associates with the spindle in mitosis like its animal homologs (Xu et al. 2007). Although its presence in the NSK has not been investigated yet, it should be very interesting to check whether this protein is also a part of the contractile endoskeleton in plants.

Three NuMA homologs of 210, 220 and 230 kDa with different solubilities were identified in the onion endoskeleton by their crossreactivity with human and *Xenopus* NuMA (Fig. 2d; Yu and Moreno Díaz de la Espina 1999). They show the same distribution as vertebrate NuMAs in an intranuclear network (excluding the nucleolus), and in the spindle matrix during mitosis which probably contributes to the stabilization of the MT bundles. Immunolabelling on resinless sections demonstrate that the protein is associated with the filaments of the endoskeleton (Yu and Moreno Díaz de la Espina 1999), that along with the potential of NuMA oligomers to form higher-order structures would indicate a structural role for NuMA in plant nuclear organization.

MAP190 is a MT- and actin-binding protein, involved in both nuclear and spindle organization in plants. MAP190 has been isolated from tobacco BY2 cells (Igarashi et al. 2000) and the gene is conserved in *Arabidopsis*. A MAP190 predicted sequence does not contain any actin filament- or microtubule (MT)-binding domains. It has an N-terminal ER-membrane retention signal; a C-terminal calmodulin-like domain; a RED repeat, essential to forming nuclear aggregates; three GHKAEQQY repeats; and several NLS and phosphorylation sites (Hussey et al. 2002; Igarashi et al. 2000). MAP190 shares some functional and distribution features with NuMA, though their sequences and secondary structures are different. It localizes in small dots in the nucleus, similar to nesprin-like proteins, NuMA and actin. After NE breakdown, MAP190 associates with the spindle and the daughter nuclei after mitosis. It appears to be involved in nuclear formation and maintenance of spindle MTs (like skeletor) and also in the cross binding of actin and MTs in the phragmoplast (Hussey et al. 2002).

The nuclear proteomes of *Arabidopsis* and rice lack actin, which has been revealed only in chickpea (Pandey et al. 2006). Besides, plants lack orthologs of the main structural nuclear ABPs (see Sect. 1.1.2) although they have functional lamin homologs and NMI was detected in the nucleolar proteome of *Arabidopsis* (Pendle et al. 2005). In contrast with the results of the proteomic analysis, conformation-specific antibodies for nuclear actin forms revealed the presence of actin forms with different solubility in isolated onion nuclei, one of them tightly bound to the NSK. Actin associates with the filaments of the onion nucleoskeleton, with a different distribution from that at transcription sites suggesting a structural role in the plant NSK (Fig. 2c; Cruz et al. 2007).

Another protein of the endoskeleton is the rice ankyrin3 homolog NALP1 that contains three ankyrin repeats and a Gly/Arg-rich domain at the N-terminus, but no other known domains or motifs. OsNALP1 and its counterpart AtAnkyrin3 are much smaller (28 and 27 kDa) than human Ankyrin3 (480 and 270 kDa). This suggests that they are probably plant specific proteins of the NSK containing the ankyrin motif (Moriguchi et al. 2005).

NMP1 is a plant specific highly conserved 36-kDa protein with a predominant α-helical structure with multiple stretches of short amphipathic regions. The protein localizes to the cytoplasm and nucleus, the latter fraction being associated with the internal NSK (Rose et al. 2003).

Several specific DNA- and MAR-binding proteins have been described in association with the plant NSK. Some of them lack coiled-coil domains and contain AT-hook motifs such as OsAHP1 (Moriguchi et al. 2005), AtAHL1 (Fujimoto et al. 2004), AHM1 (Morisawa et al. 2000), etc. These are probably not constitutive structural elements of the NSK but are proposed to be involved in the organization and attachment of chromatin fibres to this structure.

MFP1 is a conserved plant-specific, long coiled-coil protein with non-specific DNA binding activity and a dual localization in the nucleus and chloroplast (Samaniego et al. 2006). The protein consists of an N-terminus containing two conserved hydrophobic domains, a central coiled-coil rod domain and a C-terminus with a terminal DNA-binding domain and an NLS. It contains several conserved CK2 motifs (Meier et al. 1996). AcMFP1 is a basic phosphoprotein that distributes in the border zone between the condensed chromatin and ICDs and also in a new category of nuclear bodies. Its association with the NSK is regulated by CK2. Its distribution, expression and the modulation of its binding to the NSK suggest that MFP1 is not likely a basic component of the expansible nucleoskeleton but more probably is involved in chromatin binding to this structure (Fig. 2f; Samaniego et al. 2006).

The interchromatin granule network is a differentiated domain of the plant endoskeleton that would provide physical support for the organization of the multimeric complexes involved in splicing. Their specific structural protein p105 is also a component of this structure (Moreno Díaz de la Espina 1995) although its association to the NSK and functionality are unknown.

2.4 The Nucleolar Skeleton

The plant nucleolus shows a high stability and maintains its subdomain organization after isolation, suggesting that it is a self-assembling organelle (Fig. 3a,b). The isolated NoSK also maintains an organized structure after fractionation supporting the existence of an underlying structure (Fig. 3c; Novillo et al., unpublished results).

The proteome of the *Arabidopsis* NSK is enriched in nucleolar residual proteins such as fibrillarin, nucleolin, Nop 56 and Nop 58, etc. (Calikowski et al. 2003), and the plant NSK contains a prominent differentiated nucleolar subdomain in close contact with the endoskeleton (Moreno Díaz de la Espina 1995) with a different ultrastructure from that of the lamina and the endoskeleton (Fig. 1a,b). Its fibrils display a denser organization than those in the endoskeleton network and lack associated granules (Figs. 1b and 3c). The stability of the nucleolar domain of the NSK in plants differs from that in animal cells, as it survives RNase digestion prior to high salt extraction and does not depend on disulphide bond stabilization (Moreno Díaz de la Espina 1995).

The polypeptide pattern of the NoSK is complex and different from that of the nucleolus and the NSK. It displays up to 127 different polypeptide spots after silver-staining demonstrating a complex rather than a simple skeletal organization, and is enriched in acidic proteins most of which are not major nucleolar components. The NoSK shares a subset of 23 protein spots with the nucleolus and 18 with the NSK (Moreno Díaz de la Espina 1995). These proteins account for a high proportion of the NSK proteins, as the NoSK is a massive domain compared with the lamina and the endoskeleton (Calikowski et al. 2003).

The only nucleoskeletal protein so far detected in the plant NoSK is actin. It shows a diffuse distribution enriched at its periphery, different from that at nucleolar transcription foci in intact nucleoli (Fig. 3e; Cruz et al. 2007) and would have a structural role in the NoSK.

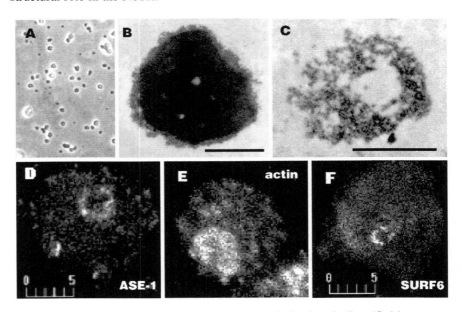

Fig. 3 Organization of the onion nucleolar skeleton. **a**: Isolated nucleoli purified in a sucrose gradient. **b, c**: EM images of an isolated nucleolus (**b**) and NoSK (**c**). The latter shows a looser fibrillar organization than the isolated nucleolus. **d–f**: Confocal immunofluorescent sections of isolated NSKs displaying the peripheral localization of some components of the NoSK. **d**: ASE-1, **e**: Actin, **f**: SURF-6. *EM bars* = 1 μm. *Confocal microscopy bars* = 5 μm

The major nucleolar proteins nucleolin and fibrillarin are enriched in the plant NoSK (Calikowski et al. 2003; Mínguez and Moreno Díaz de la Espina 1996; Moreno Díaz de la Espina and Mínguez 1996). Nucleolin is an essential protein with three structural multifunctional domains: an acidic N-terminus with several phosphorylation sites, a central region with RNA binding domains and a C-terminal GAR domain. Besides its essential roles in the different steps of ribosome biogenesis, nucleolin also controls nucleolar organization (Pontvianne et al. 2007). Fibrillarin is a highly conserved nucleolar methyltransferase involved in rRNA methylation that associates with snoRNAs and is a component of the rRNA modification and assembly complexes. Fibrillarin is a NoSK component also involved in the recruitment of rRNA processing complexes by PNBs in the NSK (Moreno Díaz de la Espina and Mínguez 1996).

Other putative proteins of the plant NoSK are MAR-binding proteins with multifunctional roles in rRNA processing and chromatin organization such as NtMARBP61 (Fujiwara et al. 2002) and pea MARBP1 and MARBP2 (Hatton and Grey 1999).

Several pol I-associated proteins have been identified by immunolabelling in the plant NoSK such as ASE-1 (Fig. 3d), the architectural transcription factor UBF, and SURF-6 (Fig. 3f) (see Sect. 1.1.3), and NO-76 that forms a peripheral scaffold in the nucleolus (Corben et al. 1989).

3 Concluding Remarks

The long controversy about the existence of an underlying NSK that spatially organizes the nuclear metabolism has been resolved in the last few years mainly by its in vivo observation through dynamic fluorescence techniques, the discovery of expandable proteins that form the contractile NSK and the determination of many architectural partners between the proteins of the NSK. The present evidence suggests that lamins and other structural proteins are linked by a variety of bridging proteins and molecular mechanisms that provide the dynamic structural framework for the organization of nuclear function that constitutes the NSK.

The architectural organization of the NSK in plants is well established and similar to that in animals. In contrast with that there is little information about its composition. The plant genomes lack orthologs of the genes that codify for some of its main structural proteins in animal cells, such as lamins and LAPs, nesprins, skeletor and NuMA, although in some of these cases they have immunologically related proteins.

Plants have functional homologs of lamins such as NIFs. Also the ability of transformed plant cells to direct the LBR to the NE and the presence of proteins that associate weakly with it demonstrate that the sequences and mechanisms involved in its targeting are conserved in plants. Plants probably have developed distant homologs of lamins and LAPs that would contain conserved essential domains but also other plant specific ones, and would be functional rather than structural homologs of lamins, as occurs with other nuclear proteins such as coilin, the structural protein of CBs. Actin is present in the plant NSK but not the main structural ABP lamins, emerin and nesprins that would bridge actin with other structural proteins in the NSK.

Besides that, plants seem to have evolved unique IF-like nuclear proteins such as NMCP1, MFP1 and NMP1, to solve the problems of nuclear architectural organization and function. The recent discovery of the implications of NMCP1-like proteins of *Arabidopsis* in nuclear organization and chromatin packaging opens the door to the investigation of the mechanisms of nuclear organization in plants.

In the future, the identification and characterization of more plant-specific nucleoskeletal proteins would clarify the molecular architecture of the plant NSK, and determine the characteristics and functions of the plant-specific subsets of proteins forming the lamina and endoskeleton. The use of mutants of NSK proteins would be also helpful in the determination of the functions of these proteins in NSK organization and function.

Acknowledgements I thank my former and present students who collaborated in collection of the data concerning the onion nucleoskeleton: A. Mínguez, W. Yu, R. Samaniego, F. Novillo, H. Li, J.R. Cruz and C. Pérez-Munive, and M. Carnota for expert technical assistance. This work was supported by Spanish DGI project BFU 2007–60142/BFI.

References

Abad PC, Mian IS, Plachot C, Nelpurackal A, Bator-Kelly C, Lelievre SA (2004) The C-terminus of the nuclear protein NuMA: phylogenetic distribution and structure. Protein Sci 13:2573–2577

Abad PC, Lewis J, Mian IS, Knowles DW, Sturgis J, Badve S, Xie J, Lelievre SA (2007) NuMA influences higher order chromatin organization in human mammary epithelium. Mol Biol Cell 18:348–361

Albiez H, Cremer M, Tiberi C, Vecchio L, Schermelleh L, Dittrich S, Kúpper K, Joffe B, Thormeyer T, von Hase J, Yang S, Rohr K (2006) Chromatin domains and the interchromatin compartment form structurally defined and functionally interacting networks. Chrom Res 14:707–733

Barboro P, D'Arrigo C, Diaspro A, Mormino M, Alberti I, Parodi S, Patrone E, Balbi C (2002) Unravelling the organization of the internal nuclear matrix: RNA-dependent anchoring of NuMA to a lamin scaffold. Exp Cell Res 279:202–218

Barboro P, D'Arrigo C, Mormino M, Coradeghini R, Parodi S, Patrone E, Balbi C (2003) An intranuclear frame for chromatin compartmentalization and higher-order folding. J Cell Biochem 88:113–120

Bettinger BT, Gilbert DM, Amberg DC (2004) Actin up in the nucleus. Nat Rev Mol Cell Biol 5:410–415

Blessing CA, Ugrinova GT, Goodson HV (2004) Actin and ARPs: action in the nucleus. Trends Cell Biol 14:435–442

Blumenthal SSD, Clark GB, Roux SJ (2004) Biochemical and immunological characterization of pea nuclear intermediate filament proteins. Planta 218:965–975

Bowman SK, Neumüller RA, Novatchkova M, Du Q, Knoblich JA (2006) The Drosophila NuMA homolog Mud regulates spindle orientation in asymmetric division. Dev Cell 10:731–742

Bridger JM, Foeger N, Kill IR, Herrmann H (2007) The nuclear lamina. Both a structural framework and a platform for genome organization. FEBS J 274:1354–1361

Calikowski TT, Meulia T, Meier I (2003) A proteomic study of the *Arabidopsis* nuclear matrix. J Cell Biochem 218:361–378

Chakalova L, Debrand E, Mitchell JA, Osborne CS, Fraser P (2005) Replication and transcription: shaping the landscape of the genome. Nat Rev Genet 6:669–677

Chen M, Shen X (2007) Nuclear actin and actin-related proteins in chromatin dynamics. Curr Opin Cell Biol 19:326–330

Chen HM, Zhou J, Dai YR (2000) Cleavage of lamin-like proteins in in vivo and in vitro apoptosis of tobacco protoplasts induced by heat shock. FEBS Lett 480:165–168

Collier S, Pendle A, Boudonck K, van Rij T, Dolan L, Shaw P (2006) A distant coilin homologue is required for the formation of Cajal bodies in Arabidopsis. Mol Biol Cell 17:2942–2951

Corben E, Butcher G, Hutchings A, Wells B, Roberts K (1989) A nucleolar matrix protein from carrot cells identified by a monoclonal antibody. Eur J Cell Biol 50:353–359

Cremer T, Cremer M, Dietzel S, Müller S, Solovei I, Fakan S (2006) Chromosome territories – a functional nuclear landscape. Curr Opin Cell Biol 18:307–316

Cruz JR, De la Torre C, Moreno Díaz de la Espina S (2007) Nuclear actin in plants. Cell Biol Int, doi:10.1016/j.cellbi.2007.11.004

De Ruijter N, Ketelaar T, Blumenthal SSD, Emons A, Schel JHN (2000) Spectrin-like proteins in plant nuclei. Cell Biol Int 24:427–438

Dittmer TA, Stacey NJ, Sugimoto-Shirasu K, Richards EJ (2007) LITTLE NUCLEI genes affecting nuclear morphology in *Arabidopsis thaliana*. Plant Cell 19:2793–2803

Dorner D, Gotzmann J, Foisner R (2007) Nucleoplasmic lamins and their interaction partners, LAP2α, Rb, and BAF, in transcriptional regulation. FEBS J 274:1362–1373

Du Q, Taylor L, Compton DA, Macara IG (2002) LGN blocks the ability of NuMA to bind and stabilize microtubules: a mechanism for mitotic spindle assembly regulation. Curr Biol 12:1928–1933

Espada J, Ballestar E, Santoro R, Fraga MF, Villar-Garea A, Nemeth A, López-Serra L, Ropero S, Aranda A, Orozco H, Moreno V, Juarranz A, Stockert JC, Längst G, Grummt I, Bickmore W, Esteller M (2007) Epigenetic disruption of ribosomal RNA genes and nucleolar architecture in DMA methyltransferase 1 (Dnmt1) deficient cells. Nucleic Acids Res 35:2191–2198

Fabian L, Xia X, Venkitaramani DV, Johansen KM, Johansen J, Andrew DJ, Forer A (2007) Titin in insect spermatocyte spindle fibers associates with microtubules, actin, myosin, and the matrix proteins skeletor, megator and chromator. J Cell Sci 120:2190–2204

Fakan S (2004) The functional architecture of the nucleus as analysed by ultrastructural cytochemistry. Histochem Cell Biol 122:83–93

Fujimoto S, Matsunaga S, Yonemura M, Uchiyama S, Azuma T, Fukui K (2004) Identification of a novel plant MAR DNA binding protein localized on chromosomal surfaces. Plant Mol Biol 56:225–239

Fujiwara S, Matsuda N, Sato T, Sonobe S, Maeshima M (2002) Molecular properties of a matrix attachment region-binding protein located in the nucleoli of tobacco cells. Plant Cell Physiol 43:1558–1567

Gonda K, Wudel J, Nelson D, Katoku-Kikyo N, Reed P, Tamada H, Kikyo N (2006) Requirement of the protein B23 for nucleolar disassembly induced by the FRGY2a family proteins. J Biol Chem 281:8153–8160

Goztmann J, Eger A, Meissner M, Grimm R, Gerner C, Sauermann G, Foisner R (1997) Two-dimensional electrophoresis reveals a nuclear-matrix associated nucleolin complex of basic isoelectric point. Electrophoresis 18:2645–2653

Graumann K, Irons SL, Runions J, Evans DE (2007) Retention and mobility of mamalian lamin B receptor in the plant nuclear envelope. Biol Cell 99:553–562

Gruenbaum Y, Margalit A, Goldman RD, Shumaker DK, Wilson KL (2005) The nuclear lamina comes of age. Nat Rev Mol Biol 6:21–31

Harborth J, Wang J, Gueth-Hallonet C, Weber K, Osborn M (1999) Self assembly of NuMA: multiarm oligomers as structural units of a nuclear lattice. EMBO J 18:1689–1700

Harborth J, Weber K, Osborn M (2000) GAS41, a highly conserved protein in eukaryotic nuclei, binds to NuMA. J Biol Chem 275:31979–31985

Hatton D, Gray JC (1999) Two MAR DNA-binding proteins of the pea nuclear matrix identify a
    new class of DNA-binding proteins. Plant J 18:417–429
He D, Nickerson JA, Penman P (1990) Coref ilamentso ft hen uclearm atrix. JC ellB iol 110:569–580
Heun P (2007) SUMOrganization of the nucleus. Curr Opin Cell Biol 19:1–6
Holaska JM, Kowalski AK, Wilson KL (2004) Emerin caps the pointed end of actin filaments:
    evidence for an actin cortical network at the nuclear inner membrane. PLoS Biol 2:1354–1362
Houben B, Ramaekers FCS, Snoeckx LHEH, Broers JLV (2006) Role of nuclear lamina-cytoskele-
    ton interactions in the maintenance of cellular strength. Biochim Biophys Acta 1773:663–674
Hussey PJ, Hawkins TJ, Igarashi H, Kaloriti D, Smertenko A (2002) The plant cytoskeleton:
    recent advances in the study of the plant microtubule-associated proteins MAP-65, MAP-190
    and the *Xenopus* MAP215-like protein, MOR1. Plant Mol Biol 50:915–924
Igarashi H, Ori H, Mori H, Shimmen T, Sonobe S (2000) Isolation of a novel 190-kDa protein
    from tobacco BY2 cells: possible involvement in the interaction between actin filaments and
    microtubules. Plant Cell Physiol 41:920–931
Jackson DA, Cook PR (1988) Visualization of a filamentous nucleoskeleton with a 23-nm axial
    repeat. EMBO J 7:3667–3677
Jockusch BM, Schoenenberger CA, Stetefeld J, Aebi U (2006) Tracking down the different forms
    of nuclear actin. Trends Cell Biol 16:391–396
Kiseleva E, Drummond SP, Goldberg MW, Rutherford SA, Allen TD, Wilson KL (2004) Actin-
    and protein-4.1-containing filaments link nuclear pore complexes to subnuclear organelles in
    Xenopus oocyte nuclei. J Cell Sci 117:2481–2490
Krauss SW, Chen C, Penman S, Heald R (2003) Nuclear actin and protein 4.1: essentials interactions
    during nuclear assembly in vitro. Proc Natl Acad Sci USA 100:10752–10757
Kruger T, Zentgraf H, Scheer U (2007) Intranuclear sites of ribosome biogenesis defined by the
    localization of early binding ribosomal proteins. J Cell Biol 177:573–578
Lam YW, Fox AH, Leung AKL, Andersen JS, Mann M, Lamond AI (2004) Proteomics of the
    nucleolus. In: Olson MOJ (ed) The nucleolus. Kluwer/Plenum, New York, pp 302–314
Ma N, Matsunaga S, Takata H, Ono-Maniwa R, Uchiyama S, Fukui K (2007) Nucleolin functions
    in nucleolus formation and chromosome congression. J Cell Sci 120:2091–2105
Magoulas C, Zatsepìna O, Jordan PWH, Jordan G, Freíd M (1998) The SURF-6 protein is a
    component of the nucleolar matrix and has a high binding capacity for nucleic acids in vitro.
    Eur J Cell Biol 75:174–183
Margalit A, Brachner A, Gotzmann J, Foisner R, Gruenbaum Y (2007) Barrier-to-autointegration
    factor – a BAFfling little protein. Trends Cell Biol 17:202–208
Masuda K, Takahashi S, Nomura K, Arimoto M, Inoue M (1993) Residual structure and constituent
    proteins of the peripheral framework of the cell nucleus in somatic embryos from *Daucus
    carota* L. Planta 191:523–540
Masuda K, Xu ZJ, Takahasi S, Ito A, Ono M, Nomura K, Inoue M (1997) Peripheral framework
    of carrot cell nucleus contains a novel protein predicted to exhibit a long α-helical domain.
    Exp Cell Res 232:173–181
Mattagajasingh SN, Huang SC, Harternstein J, Snyder M, Marchesi VT, Benz EJ Jr (1999) A non
    erythroid isoform of protein 4.1R interacts with the nuclear mitotic apparatus (NuMA) protein.
    J Cell Biol 145:29–43
McDonald D, Carrero G, Andrin C, de Vries G, Hendzel MJ (2006) Nucleoplasmic β-actin exists
    in a dynamic equilibrium between low-mobility polymeric species and rapidly diffusing
    populations. J Cell Biol 172:541–552
McNulty AK, Saunders MJ (1992) Purification and immunological detection of pea nuclear
    intermediate filaments: evidence for plant nuclear lamins. J Cell Sci 103:407–414
Meagher RB, Fechheimer M (2003) The *Arabidopsis* cytoskeletal genome. In: Somerville CR,
    Meyerowitz EM (eds) The arabidopsis book. American Society of Plant Biologists, Rockville,
    MD, pp 1–26, doi:10.119/tab.0096
Meier I, Phelan T, Gruissem W, Spiker S, Schneider D (1996) MFP1, a novel plant filament-like
    protein with affinity for matrix attachment region DNA. Plant Cell 8:2105–2115

Melcer S, Gruenbaum Y, Krhone G (2007) Invertebrate lamins. Exp Cell Res 313:2157–2166

Mínguez A, Moreno Díaz de la Espina S (1993) Inmunological characterization of lamins in the nuclear matrix of onion cells. J Cell Sci 106:431–439

Mínguez A, Moreno Díaz de la Espina S (1996) In situ localization of nucleolin in the plant nuclear matrix. Exp Cell Res 222:171–178

Moreno Diaz de la Espina S (1995) Nuclear matrix isolated from plant cells. In: Berezney R, Jeon KW (eds) Int Rev Cytol, vol. 162B. Academic, San Diego, pp 75–139

Moreno Díaz de la Espina S, Mínguez A (1996) Post-mitotic assembly of the nucleolus. I. The internal matrix network is a recruitment site for processing nucleolar components in prenucleolar bodies. Chrom Res 4:103–110

Moriguchi K, Suzuki T, Ito Y, Yamazaki Y, Niwa Y, Kurata N (2005) Functional isolation of novel nuclear proteins showing a variety of subnuclear localizations. Plant Cell 17:389–403

Morisawa G, Han-Yama A, Moda I, Tamai A, Iwabuchi M, Meshi T (2000) AHM1 a novel type of nuclear matrix localized, MAR binding protein with a single AT hook and a J domain-homologous region. Plant Cell 12:1903–1916

Naetar N, Hutter S, Dorner D, Dechat T, Korbei B, Gotzman J, Beug H, Foisner R (2007) LAP2α-binding protein LINT-25 is a novel chromatin-associated protein involved in cell cycle exit. J Cell Sci 120:737–747

Nalepa G, Harper JW (2004) Visualization of a highly organized intranuclear network of filaments in living mammalian cells. Cell Motil Cytoskel 59:94–108

Nickerson JS (2001) Experimental observations of a nuclear matrix. J Cell Sci 114:463–474

Pandey A, Choudhary MK, Bhushan D, Chattopadhyay A, Chakraborty S, Datta A, Chakraborty N (2006) The nuclear proteome of chickpea (Cicer arietinum L.) reveals predicted and unexpected proteins. J Proteome Res. 5:3301–3311

Pendle AF, Clark GP, Boon R, Lewandoska D, Lam YW, Andersen J, Mann M, Lamond AI, Brown JWS, Shaw P (2005) Proteomic analysis of the Arabidopsis nucleolus suggests novel nucleolar functions. Mol Biol Cell 16:260–269

Pliss A, Koberna K, Vecerova J, Malinsky J, Masata M, Fialova M, Raska I, Berezney R (2005) Spatio-temporal dynamics at rDNA foci: global switching between DNA replication and transcription. J Cell Biochem 94:554–565

Polzikov M, Zatsepina O, Magoulas C (2005) Identification of an evolutionary conserved SURF-6 domain in a family of nucleolar proteins extending from human to yeast. Biochem Biophys Res Comm 327:143–149

Pontvianne F, Matía I, Douet J, Tourmente S, Medina FJ, Echeverría M, Saéz-Vázquez J (2007) Characterization of AtNUC-L1 reveals a central role of nucleolin in nucleolus organization and silencing of AtNUC-L2 gene in Arabidopsis. Mol Biol Cell 18:369–379

Prieto JL, McStay B (2007) Recruitment of factors linking transcription and processing of pre-rRNA to NOR chromatin is UBF-dependent and occurs independent of transcription in human cells. Genes Dev 21:2041–2054

Qi H, Rath U, Wang D, Xu YZ, Ding Y, Zhang W, Blacketer MJ, Paddy MR, Girton J, Johansen J, Johansen KM (2004) Megator, an essential coiled-coil protein that localizes to the putative spindle matrix during mitosis in Drosophila. Mol Biol Cell 15:4854–4865

Qi H, Rath U, Wang D, Xu YZ, Ding Y, Blacketer MJ, Girton J, Johansen J, Johansen KM (2005) EAST interacts with megator and localizes to the putative spindle matrix during mitosis in Drosophila. J Cell Biochem 95:1284–1291

Rath U, Wang D, Ding Y, Xu YZ, Qi H, Blacketer MJ, Girton J, Johansen J, Johansen KM (2004) Chromator a novel and essential chromodomain protein interacts directly with the putative spindle matrix protein skeletor. J Cell Biochem 93:1033–1047

Rath U, Ding Y, Deng H, Qi H, Bao X, Zhang W, Girton J, Johansen J, Johansen KM (2006) The chromodomain protein chromator interacts with JIL-1 kinase and regulates the structure of Drosophila polytene chromosomes. J Cell Sci 119:2332–2341

Rose A, Gindullis F, Meier I (2003) A novel alpha-helical protein, specific to and highly conserved in plants, is associated with the nuclear matrix. J Exp Bot 54:1133–1141

Samaniego R, Jeong SY, De la Torre C, Meier I, Moreno Díaz de la Espina S (2006) CK2 phosphorylation weakens 90-kDa MFP1 association to the nuclear matrix in *Allium cepa*. J Exp Bot 57:101–111

Schirmer EC, Foisner R (2007) Proteins that associate with lamins: many faces, many functions. Exp Cell Res 313:2167–2169

Schonenberger CA, Buchmeier S, Boerries M, Sütterlin R, Aebi U, Jockusch BM (2005) Conformation-specific antibodies reveal distinct actin structures in the nucleus and the cytoplasm. J Struct Biol 152:157–168

Shaklai S, Amariglio N, Rechavi G, Simon AJ (2007) Gene silencing at the nuclear periphery. FEBS J 274:1383–1392

Shaw P, Doonan J (2005) The nucleolus. Playing by different rules? Cell Cycle 4:102–105

Shumaker DK, Kuczmarski ER, Goldman RD (2003) The nucleoskeleton: lamins and actin are major players in essential nuclear functions. Curr Opin Cell Biol 15:358–366

Sone M, Hayashi T, Tarui H, Agata K, Takeichi M, Nakagawa S (2007) The mRNA-like noncoding RNA Gomafu constitutes a novel domain in a subset of neurons. J Cell Sci 120:2498–2506

Suzuki S, Fujiwara T, Kanno M (2007) Nucleolar protein NOP25 is involved in nucleolar architecture. Biochem Biophys Res Comm 358:1114–1119

Tzur YB, Wilson KL, Gruenbaum Y (2006) SUN-domain proteins: "Velcro" that links the nucleoskeleton to the cytoskeleton. Nat Rev Mol Cell Biol 7:782–788

Uzbekov R, Timirbulatova E, Watrin E, Cubizolles F, Ogereau D, Gulak P, Legagneux V, Polyakov WJ, Le Guellec K, Kireev I (2003) Nucleolar association of pEg7 and XCAP-E, two members of the *Xenopus laevis* condensin complex in interphase cells. J Cell Sci 116:1667–1678

Vlcek S, Foisner R (2007) A-type lamin networks in light of laminopathic diseases. Biochim Biophys Acta 1773:661–674

Walker DL, Wang D, Jin Y, Wang Y, Johansen J, Johansen KM (2000) Skeletor, a novel chromosomal protein that redistributes during mitosis provides evidence for the formation of a spindle matrix. J Cell Biol 151:1401–1411

Wasser M, Chia W (2000) The EAST protein of *Drosophila* controls an expandable nuclear endoskeleton. Nat Cell Biol 2:268–275

Wasser M, Chia W (2005) EAST interacts with Megator and localizes to the putative spindle matrix during mitosis in Drosophila. J Cell Biochem 95:1284–1291

Xu XM, Rose A, Muthuswamy S, Jeong SY, Venkatakrishnan S, Zhao Q, Meier I (2007) NUCLEAR PORE ANCHOR, the *Arabidopsis* homolog of Tpr/Mlp1/Mlp2/Megator, is involved in mRNA export and SUMO homeostasis and affects divers aspects of plant development. Plant Cell 19:1537–1548

Ye J, Zhao J, Hoffmann-Rohrer U, Grummt I (2008) Nuclear myosin I acts in concert with polymeric actin to drive RNA polymerase I transcription. Genes Dev 22:322–330

Yu W, Moreno Díaz de la Espina S (1999) The plant nucleoskeleton: ultrastructural organization and identification of NuMA homologues in the nuclear matrix and mitotic spindle of plant cells. Exp Cell Res 246:516–526

Zastrow MS, Flaherty DB, Venían GM, Wilson KL (2006) Nuclear Titin interacts with A- and B-type lamins in vitro and in vivo. J Cell Sci 119:239–249

Zhang Q, Ragnauth CD, Skepper JN, Worth NF, Warren DT, Roberts RG, Weissberg PL, Ellis JA, Shanahan CM (2005) Nesprin-2 is a multiisomeric protein that binds lamin and emerin at the nuclear envelope and forms a subcellular network in skeletal muscle. J Cell Sci 118:673–687

Zheng Y, Tsai MY (2006) The mitotic spindle matrix. A fibro-membranous lamin connection. Cell Cycle 5:2345–2347

The Role of Nuclear Matrix Attachment Regions in Plants

George C. Allen

Abstract Regions of DNA that bind to the nuclear matrix, or nucleoskeleton, are known as Matrix Attachment Regions (MARs). MARs are thought to play an important role in higher-order structure and chromatin organization within the nucleus. MARs are also thought to act as boundaries of chromosomal domains that act to separate regions of gene-rich, decondensed euchromatin from highly repetitive, condensed heterochromatin. Herein I will present evidence that MARs do indeed act as domain boundaries and can prevent the spread of silencing into active genes. Many fundamental questions remain unanswered about how MARs function in the nucleus. New findings in epigenetics indicate that MARs may also play an important role in the organization of genes and the eventual transport of their mRNAs through the nuclear pore.

1   Introduction

For many years it has been known that the expression of a transgene in eukaryotic organisms can vary widely between independent transformants (Allen et al. 1988). The unpredictable, varied expression found in transgenic plants has been called "chromosomal position effects," and is attributed to the characteristics of the site of integration (Alberts and Sternglanz 1990; Dean et al. 1988; Nagy et al. 1985). Chromosomal position effects are primarily due to transgene integration events that can occur within euchromatin, which contains the majority of expressed genes, or condensed chromatin, such as heterochromatin (reviewed in Taddei et al. 2004). The pre-existing chromatin structure at the site of integration ultimately determines the transgene expression level, either negatively (silenced) or positively (enhanced). Transgene expression can be highly variable since transgene integration cannot yet

G.C. Allen
North Carolina State University, Department of Horticultural Science, 1203 Partners 2, Campus Box 7550, Raleigh, NC 27606-7550, USA
e-mail: george_allen@ncsu.edu

Plant Cell Monogr, doi:10.1007/7089_2008_21
© Springer-Verlag Berlin Heidelberg 2008

be easily targeted to a specific genomic location with a favorable chromatin structure.

There are many examples that complicate both the use of transgenes as research tools and their application in plant-improvement programs (De Neve et al. 1999; Gepts 2002; James et al. 2004). Variability due to silencing and related processes requires the evaluation of many transformation "events," or independent transformants, in order to interpret the phenotypic effects of transgenes. This, in turn, increases the cost of labor and cause for regulatory concern. Because silencing can occur spontaneously in later generations of lines initially showing good expression, selecting stably expressing lines from a population of independent transformants, or even from the T1 generation, does not guarantee stability of expression over subsequent generations (Bourdon et al. 2002; Bregitzer and Tonks 2003; Chareonpornwattana et al. 1999; Levin et al. 2005; Vain et al. 2002). In order to avoid major problems, a number of candidate lines must be maintained through multiple generations before it is possible to select those showing stable and predictable expression.

In this chapter I explore ideas why MARs[1] improve the transformation frequency and reliability of transgene expression in some cases but seem to have no effect in other cases. I discuss practical strategies for preventing undesirable gene silencing and provide examples showing that MARs are, counterintuitively, beneficial for enhancing transgene RNA silencing (RNAi) when a gene knockout (knockdown) is desired. I also describe the other critical roles that MARs play in regulating nuclear structure and genome organization, beyond their role in gene expression.

In order to understand several controversial ideas regarding MARs it is important to understand the nuclear matrix, which is also known as the nuclear scaffold, or nucleoskeleton. A detailed description of the nucleoskeleton is provided in an accompanying chapter in this book by Moreno Diaz de la Espina. It is thought that the effect of MARs on gene expression is achieved through their impact on chromatin structure. I briefly describe the role of histone modifications with particular emphasis upon the replacement histone H2A.Z. Increasingly, such histone variants are being recognized as mediators that can act in concert with boundary elements to separate active from inactive chromatin. The reader is also referred to accompanying chapters in this book that describe work on nuclear pores (A. Rose) and on chromatin domains (P. Fransz).

---

[1] It is important to note that during the past twenty 20 plus years various nomenclature have been used for MARs which include Scaffold Attachment Region (SAR) (Allen et al. 1996; Bode and Maass 1988), Boundary Elements (Cuvier et al. 2002; Mlynarova et al. 2003; Pathak et al. 2007), certain types of Chromatin Insulators (Istomina et al. 2003; Valenzuela and Kamakaka 2006), or a combination of both (S/MAR) (Goetze et al. 2005; Heng et al. 2004). These terms are used interchangeably throughout this chapter.

1.1 Chromosomal Loop Domains

Early studies with animal cell lines demonstrated that the "chromosomal position effect" could be prevented if the transgene was flanked by "Matrix Attachment Regions" (MARs[2]), which are DNA elements that bind to the nuclear matrix[3] (Mirkovitch et al. 1984). Electron microscopy studies from Paulson and Laemmli (1977) provided an elegant demonstration that when the core histones are removed, DNA from metaphase chromosomes becomes unpackaged and reveals higher-order structure. A series of large DNA loops is bound at their bases to a core structure, which had been identified earlier as the nuclear matrix by Berezney and Coffey (1974). When the data from higher-order structure was combined with gene expression studies, a new model, known as the loop domain model, was developed (Bodnar 1988; Cockerill and Garrard 1986; Cook 1989; Gasser and Laemmli 1986; Stief et al. 1989).

According to the loop domain model, MARs act as a barrier to the effect of surrounding chromatin (Fig. 1). When applied to transgenic plants, this model would effectively make the transgene domain independent. A major prediction of the loop domain model is that a transgene, flanked by the "protecting" MARs, remains active independently of the local chromatin state. A corollary prediction is that transgenes lacking MARs are susceptible to the impact of the surrounding chromatin structure and their expression level is dependent upon the local chromatin state. By acting independently, each transgene domain would contribute equally to transgene expression, which would increase in proportion to the number of transgenes. Thus, if more copies of a MAR-flanked transgene are integrated, a proportional increase in the expression of the transgene will result. Such predictions made from the loop

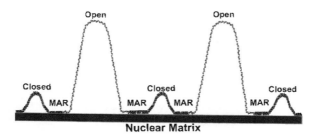

Fig. 1 Loop domain model. Active genes are within the open loops with an open chromatin structure. The open loops are bounded at their base by Matrix Attachment Regions (MARs) that bind to the nuclear matrix (*grey*). The chromatin structure at the MAR is in a transition zone that separates the open loop from the closed loops. The closed loops contain compact nucleosome

---

[2] MARs also called SARs (Scaffold Attachments Regions) or S/MARs.
[3] Nuclear matrix is also called nucleoskeleton or nuclear scaffold.

domain model were supported by the results from the early animal cell studies in which transgenes flanked by MARs from either the human β-globin gene (Grosveld et al. 1987) or the chicken lysozyme A-element (Phi-Van et al. 1990; Stief et al. 1989) resulted in transgene expression that was proportional to transgene number.

2   The Role of MARs in Higher-Order Chromatin Structure in the Plant Nucleus

2.1   Packaging the Genome

2.1.1   DNase I Sensitive Sites

Genomic DNA must be carefully folded in an organized manner in order to fit within the interphase nucleus. The packaging of chromatin is accomplished at several levels starting with the interaction of double-stranded DNA with two molecules of each of the core histones H2A, H2B, H3, and H4 to form the 11-nm chromatin fiber. The 11-nm fiber is further packaged by interacting with H1, the linker histone, to form the condensed 30-nm nucleosome. The details of the subsequent packaging steps to form the higher-order condensed structures are still largely unknown. It is generally thought that chromatin loops of up to 300 nm are bound at their base by MARs to a non-histone protein-RNA core, which is believed to be the nuclear matrix as shown in Fig. 2. When the core histones are removed, the ~300-nm loops are unpackaged. Measurements by Paulson and Laemmli (1997) estimated that these histone-free loops averaged 10–12 μm in length in *Drosophila* metaphase chromosomes, which represents an increase in length of 30–40-fold.

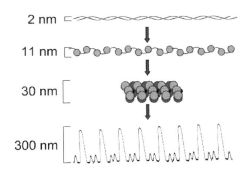

Fig. 2 Chromatin packaging. The packaging of chromatin is accomplished at several levels starting with the interaction of double-stranded DNA with two molecules of each of the core histones H2A, H2B, H3, and H4 to form the 11-nm chromatin fiber. The 11-nm fiber is further packaged by interacting with H1, the linker histone, to form the condensed 30-nm nucleosome. It is generally thought that chromatin loops of up to 300 nm are bound at their base by MARs to a non-histone protein-RNA core, which is believed to be the nuclear matrix

The different levels of chromatin structure can shorten a double-stranded molecule of DNA by up to 50,000-fold (Alberts et al. 2002).

When a gene is activated, the chromatin structure becomes less compact and the DNA becomes more accessible to DNase I (Spiker et al. 1983; Weintraub and Groudine 1976). MARs are thought to function as the boundaries of the chromatin loop domains and to demarcate active chromatin from inactive chromatin (Bode and Maass 1988; Bonifer et al. 1991; Gasser and Laemmli 1986; Gromova et al. 1995). Paul and Ferl (1998) used differences in chromatin accessibility to DNase I to compare the higher-order chromatin structure in both Arabidopsis and maize nuclei. They found that DNase I cleavage at "super-hypersensitive" sites resulted in domains that averaged 45 kb for maize and 25 kb for Arabidopsis. They proposed a model that these larger domains represent a fundamental structural loop domain bound to the nuclear matrix at "loop basements," or LBARs.

It is now well recognized that gene-coding sequences alone do not provide answers to how the genome is regulated. A new effort to understand the function of the epigenome (Crawford et al. 2006) has been initiated with the goal of understanding the function of the non-coding regions of the human genome, which includes mapping the DNase I hypersensitive sites using tiled microarrays (Giresi and Lieb 2006; Sabo et al. 2006). Because of the size of the human genome, researchers are limited to focusing on selected regions, which only encompass approximately 1% of the entire genome (Greally 2007). In contrast, Arabidopsis, with such a small genome, offers tremendous advantages for such studies. In principle, either entire chromosomes or even the entire genome could be studied in detail. Unfortunately, while the analytical capability does exist, the mapping of DNase I hypersensitive sites in Arabidopsis has to date been limited to relatively small regions.

Kodama et al. (2007) mapped DNase I hypersensitive sites in an 80-kb region from Chromosome 5 of *Arabidopsis thaliana* that contains 30 genes with expression levels that vary greatly. A total of 40 DNase I hypersensitive sites were found and all were located at the 5′ and 3′ ends of 28 of the 30 genes. These data suggest that the average size of a DNase I "hypersensitive domain" within the 80-kb region is ~2 kb. If such DNase I hypersensitive sites found by Kodama et al. (2007)represent true chromosomal domains, then the size is much smaller than the DNase I hypersensitive domains found in either animal studies or even the 25-kb estimate for *Arabidopsis* by Paul and Ferl (1998). It is important to note that the apparent difference in DNase domain size may simply be due to the use of an 80-kb region (Kodama et al. 2007) versus a 130,000-kb region (Paul and Ferl 1998). Furthermore, the Kodama et al. (2007) result seems to differ from previous studies that show that DNase I hypersensitivity varies depending upon the expression of the gene. For example, Vega-Palas and Ferl (1995) note that DNase I-hypersensitive sites in the 5′ flanking region of transcriptionally active genes are associated with the presence of transcriptional regulatory factors, which also reflects a de-condensation of the chromatin. Although more studies will certainly follow, the Kodama et al. study suggests that DNase I hypersensitivity does not reflect transcriptional activity, which is likely to be controversial.

2.1.2 MARs and Histone Modification

In eukaryotic nuclei the domains defined by differences in DNase I sensitivity are also characterized by the differences in post-translational modifications of the histones within the domain. Histone modifications within DNase I-accessible and transcriptionally active domains are typically characterized by hyper-acetylation and di- or trimethylation of the lysine-4 of H3. Conversely, tran-scriptionally inert or inactive chromatin is tightly packaged and thus insensitive to DNase I, and the lysine-9 of H3 is hypoacetylated and methylated (Johnson et al. 2002; Soppe et al. 2002; Tamaru and Selker 2001; and reviewed in Tariq and Paszkowski 2004). Whole genome analysis of *Arabidopsis* using Chromatin Immunoprecipitation (ChIP) analysis has confirmed that transcriptionally active regions are highly enriched with H3 lysine-4 di- or trimethylation, whereas inactive heterochromatic regions are enriched in H3 lysine-9 trimethylation (Gendrel et al. 2005).

As noted earlier, MARs, or boundary elements, are proposed to define chro-mosomal domains and to prevent transcriptionally active regions from being silenced by the spread of condensed heterochromatic regions (Martienssen 2003). While it is now clear that the inactive and active chromosomal domains are enriched in specific forms of modified histones, it is not known what type of histone is actually associated with MARs. The Grey lab has dissected the chromatin structure of the pea *PET* gene and the initial studies identified a downstream MAR region (Slatter et al. 1991). Further studies identified an A/T-rich region in the promoter that bound to High Mobility Group I/Y (HMG) proteins (Pwee et al. 1994). Transgenic plant studies showed that the A/T-rich region enhanced the expression of a b-glucuronidase (*GUS*) reporter gene and that when the 31-bp A/T core was included, the reporter gene expression also increased (Sandhu et al. 1998). Chua et al. (2001) then tested the pea *PetE* locus for changes in nuclease sensitivity and H3 and H4 acetylation during three dif-ferent transcriptional states (roots, etiolated shoots, and green shoots). As expected, the DNase I sensitivity increased during activation, and H3 and H4 from *PetE* were hyperacetylated in the green shoots, and were particularly enriched in the promoter/enhancer region.

A subsequent study by Chua et al. (2003) found that the A/T-rich transcriptional enhancer of the pea *Pet1* bound to the nuclear matrix when either an in vitro binding assay or in vivo enrichment in the matrix fraction in the transgenic tobacco was used. Because MARs are defined as DNA elements that bind the matrix, this enhancer was also a MAR. Transgenic tobacco transformed with an enhancer *Pet1* promoter fragment controlling *GUS* (E-*Pet1*:*GUS*) had higher *GUS* expression levels than the plants transformed with the same construct lacking the enhancer (*Pet1*:*GUS*). Differences in the H3 and H4 acetylation patterns reflected the expres-sion differences with the pea E*Pet1*:*GUS* transformants showing increased levels of H3 and H4 acetylation that spread into the 5' end of *GUS*. These data are consistent with a model for MARs being associated with open chromatin that has the increased H3 and H4 acetylation patterns expected for active genes.

2.1.3   The Transition Zone: MARs, Boundaries and Histone Variants

Studies to understand the transition, or boundary zone between inactive and active chromosomal domains have become increasingly important as intergenic, non-coding genomic regions are recognized for controlling critical processes such as transcription, DNA replication, cell division, differentiation, and development. In the popular press the genome sequence and how it is regulated has been compared to a computer; with the primary sequence referred to as the hardware and the epigenome referred to as the software. While changes in genomic sequence are generally thought to occur slowly, changes to the epigenome can occur rapidly, which may help the organism to readily adapt to environmental change.

It is now known that nearly the entire genome is transcriptionally active, which includes non-coding regions that can produce regulatory RNAs. These results were initially found using the fission yeast *Schizosaccharomyces pombe* (Hall et al. 2002; Volpe et al. 2002) and showed that centromeric heterochromatin is maintained by RNAi. The same, or a closely related mechanism is also responsible for forming (Martienssen 2003; Zilberman et al. 2004) or maintaining (Gendrel et al. 2002; May et al. 2005) heterochromatin in higher eukaryotes (reviewed by Henderson and Jacobsen 2007; Lippman and Martienssen 2004; Martienssen et al. 2005; Zaratiegui et al. 2007). However, major questions remain regarding the composition and regulation of genomic boundary regions and how they function to separate the active from inactive chromosomal domains, preventing the spread of transcriptionally inactive, condensed chromatin into the transcriptionally active, de-condensed chromatin.

The process of how gene silencing spreads has been an enigma for many years. It is well known that the proximity of a transgene integration site to heterochromatin can result in silencing of the transgene (Assaad et al. 1993; Dorer and Henikoff 1997; and reviewed in Henikoff 1998 and Jacobsen 1999). In addition, the organization of a transgene locus also is known to play a major role, particularly when multiple copies integrate at a single site (Jorgensen et al. 1996; Stam et al. 1997; and reviewed in Allen et al. 2000 and Thompson et al. 2006). For instance, most transgene integrations that result in an inverted repeat structure become silenced and form condensed chromatin (Martienssen 2003; Muller et al. 2002; Pecinka et al. 2005). Condensed chromatin is thought to be able to spread by a self-propagated RNAi-mediated process, unless it is physically blocked by either a bound protein or a change in the chromatin structure (Bi et al. 2004; and reviewed in Richards and Elgin 2002; Talbert and Henikoff (2006); and Gaszner and Felsenfeld 2006).

Recent studies in budding yeast (*Saccharomyces cerevisiae*) have shown that the histone variant H2A.Z is found at the transition zones between active and inactive chromatin (Meneghini et al. 2003). Nucleosomes that include both histone variants H2A.Z (H1Z in *S. cerevisiae*), and another variant, H3.3, are more susceptible to disruption (Jin and Felsenfeld 2007), leading to an increase in local chromatin accessibility. Of particular interest is the finding by Meneghini et al. (2003) that the interaction of H2A.Z with boundary elements can act synergistically to block the spread of heterochromatin. To date, the only studies to demonstrate the H2A.Z-insulator block have been done in *S. cerevisiae*. While the role of H2A.Z is less clear in higher eukaryotes (Wong et al. 2007), it is necessary for viability in *Tetrahymena thermophila*, *Drosophila melanogaster*, and mice (Sarcinella et al. 2007).

Arabidopsis thaliana H2A.Z (H2A.F/Z) is highly conserved (Kamakaka and Biggins 2005; Meneghini et al. 2003) and appeared very early in evolution (Callard and Mazzolini 1997). This raises the possibility that Arabidopsis H2A.Z may have retained similar functions, to bind to boundary elements and block the spread of heterochromatin. In yeast, H2A.Z (H1Z) is a replacement histone for H2A. The replacement is catalyzed by SWR1C, a member of the SWI2/SNF2 super-family with orthologs in Arabidopsis. In Arabidopsis, flowering in long day conditions is contingent upon the successful establishment of epigenetic silencing of *FLC*, a repressor of the transition to flowering. This is achieved by histone trimethylation at H3 lysine 27 and deacetylation and is maintained by proteins that establish a Polycomb-like chromatin regulation (Greb et al. 2007). Mutations in *FLC* cause early flowering, the same phenotype as various mutants of *Arabidopsis thaliana* with defective *SWR1C* (Choi et al. 2007). Deal et al. (2005) showed that mutations in *SUF3* (*ARP6* in yeast), which is a nuclear actin-related protein and also part of the SWR1 chromatin remodeling complex, also result in an early flowering phenotype, similar to the *FLC* mutations. Finally, studies by Deal et al. (2007) showed that *Arabidopsis thaliana* H2A.Z is required to suppress flowering by activating *FLC*. Such observations that H2A.Z potentiates transcription are similar to the proposed role of H1Z in the yeast.

2.1.4 Do MARs and Boundary Elements Associate with the Nuclear Pore Complex?

The nuclear pore allows communication between the interior of the interphase nucleus and the exterior cytoplasm, as described in detail in a separate chapter in this book (A. Rose). It is generally assumed that mRNA must be exported in order to be transcribed but given the packaging constraints of nuclear DNA, mRNA export is unlikely to occur by simple diffusion or stochastic interactions. Here I briefly describe a possible role for the nuclear pore in the localization of active genes (Aguilera 2005; Cabal et al. 2006; Casolari et al. 2004; Ishii et al. 2002; Taddei et al. 2006). In yeast, mRNA transport from an actively transcribing gene has been shown to "track" from the nuclear interior to the nuclear pore. However, data from recent studies with both yeast and *Drosophila* have also shown that chromatin redistribution or movement may contribute to the transport process. H2A.Z has been proposed to facilitate this process providing active genes access to the nuclear pore. The molecular intermediates that control this process could interact with MARs or use MARs as a type of DNA tether to shuffle parts of the chromosome into territories with access to a nuclear pore.

Data from a recent study by Brickner et al. (2007) indicate that H2A.Z can interact with promoter regions of an active gene, thereby localizing the genes to the nuclear pore. Interestingly, even recently repressed genes retain bound H2A.Z and remained associated with the nuclear pore complex (NPC). Chromatin immu-noprecipitation (ChIP) experiments have been used to show that transcriptionally active genes can occur in a complex that includes components of the nuclear pore complex, such as Nup2. Brickner et al. (2007) suggest that Nup2 may recruit or "tether" H2A.Z and its associated genes, to the nuclear pore to enable rapid re-activation (Dilworth et al. 2005; Ishii et al. 2002). Such "tethering" may promote boundary activity, thus preventing the spread of condensed chromatin into tran-scriptionally poised regions. Brickner et al. (2007) also note that previous data with mutants lacking either H2A.Z, or Nup2 show an increase in the spread of silent chromatin (Dilworth et al. 2005; Meneghini et al. 2003). Boundary elements are thought to associate with the NPC, allowing active genes to be correctly positioned for either active transcription or to be poised for transcription. Thus, H2A.Z binding allows for both proper localization of a gene to the NPC, and proper demarcation of heterochromatic silencing by its interaction with boundary elements. Furthermore, by remaining associated with recently active chromatin, H2A.Z confers epigenetic memory (Brickner et al. 2007).

One caveat that remains to be addressed is that all of the studies that have shown H2A.Z association with the nuclear pore have been done in *S. cerevisiae*. While most of the cellular processes that are found in *S. cerevisiae* also occur in higher eukaryotes, there are a few important exceptions. In addition to the small size of its genome and nucleus, post-transcriptional gene silencing appears to be lacking in *S. cerevisiae* while some form of this silencing occurs in almost all other eukaryotes, including quelling in fungi (Valenzuela and Kamakaka 2006). Additionally, homologous recombination readily occurs in *S. cerevisiae*. Although many questions remain from the *S. cerevisiae* studies, the process is thought to be highly conserved across eukaryotes (Cole and Scarcelli 2006; Ragoczy et al. 2006). Taken together as a general model, tethering of active or recently active gene regulatory regions by MAR-bound proteins, such as H2A.Z, to the NPC may both block the spread of condensed chromatin and facilitate the transport of mRNA from the nucleus. However, questions will remain until experiments are done in plants to directly test this model.

3 The Use of MARs to Prevent Transgene Silencing

3.1 MARs and Transgene Expression

Encouraging results from animal studies (Grosveld et al. 1987; Phi-Van et al. 1990; Stief et al. 1989) suggested that MAR-flanked transgenes would be an important tool for preventing chromosomal position effects in plants (Allen et al. 1993, 1996; Breyne et al. 1992; Mlynarova et al. 1994; Schoffl et al. 1993). A diversity of pro-moters and MAR sequences have been used to test the level of expression of the β-glucuronidase (*GUS*) reporter gene. Allen et al. (1993) used the *GUS* gene driven by the strong CaMV 35S promoter, which was flanked by either the yeast ARS-1[4] (Allen et al. 1993), which is known to be a MAR (Amati and Gasser 1988), or a tobacco MAR from the *Rb7-3* gene (Allen et al. 1996; Hall et al. 1991). Breyne et al. (1992) used *GUS* driven by the weaker nopaline synthase promoter and MARs from either the soybean lectin gene or the human β-globin locus to flank the transgene cassette. Mlynarova et al. (1994) used a *GUS* reporter gene driven by the promoter from apoprotein 2 of the light harvesting complex of photosystem I of the potato (*Lhca3*) promoter, which was flanked by MARs from the chicken lysozyme A element. Schoffl et al. (1993) also used a *GUS* reporter transgene, which was controlled by the heat-shock inducible promoter from the heat-shock inducible *Gmhsp17.3-B* gene from soybean. A 395-bp MAR from a different soybean heat shock gene (*Gmhsp 17.6L*) flanked the transgene cassette.

When NT1 tobacco suspension culture cell lines were transformed with the ARS-1 MAR-flanked *GUS* gene, they had 12-fold greater GUS expression when compared to the control transgenic lines with transgenes lacking MARs (Allen et al. 1996). When a plant MAR from the *RB7* gene was used, there was nearly 60-fold greater expression. An important result from both of these experiments was that transgene expression levels were not generally proportional to the transgene copy number. When the yeast MAR ARS-1 was used, expression appeared to be proportional to copy number until a "threshold" of 20 copies was reached (Allen et al. 1993) and transgene expression drastically decreased. The results were similar when the Rb7 MAR was used; lines with greater than ten copies (Allen et al. 1996) had greatly reduced expression. In contrast, when either the human β-globin MAR (Breyne et al. 1992) or the chicken lysozyme A-element MAR (Mlynarova et al. 1994) was used, a modest increase *GUS* expression was seen but the expression between transformants was less variable. In apparent contrast to the results of both groups, Schoffl et al. (1993) found that including flanking *Gmhspl 7.6-L* MARs resulted in expression levels that were both proportional to transgene copy number and five- to ninefold higher in expression following heat shock induction.

A comparison of the experimental designs used for each experiment reveals important differences. Allen et al. (1993, 1996) used microprojectile bombardment, a procedure that results in higher transgene copy numbers, and co-transformation, which in principle physically separates the reporter transgene from the selectable marker (see Sect. 3.1.1). Higher transgene copy numbers typically results in a higher probability of gene silencing. Allen et al. (1993, 1996) found that lines with genes lacking MARs showed much higher levels of silencing than the lines that contained MARs. The majority of the transformed lines that were studied also had multiple copies of 35S:*GUS* that were frequently at the same locus regardless of whether MARs were present or absent (Allen et al. 1993, 1996).

Studies during the past decade have generally concluded that the effect of flanking a transgene with MAR sequences is positive, although variation is still high. I have not included a detailed comparison of all of the MAR studies in plants

---

[4] Autonomously Replicating Sequence (ARS-1) from Saccharomyces is known to act as a SAR (MAR).

because a comprehensive comparison of these experiments has been published elsewhere by Allen et al. (2000) and more recently by Thompson et al. (2006). In the following sections, I try to add some understanding for why such apparent variation can be explained by RNAi, which is not measured when transgene mRNA abundance or protein levels are used to assess the effect of MARs. Thus, if the RNAi-mediated reduction in transgene expression were to be taken into account an entirely different interpretation may result. Increasingly it is being recognized that MARs seem to prevent gene silencing in *cis*, but do not appear to be effective on gene silencing in *trans*. With this knowledge, new strategies are being developed that combine the use of MARs for preventing or reducing RNA-mediated *cis*-silencing with new methods for reducing RNA-mediated *trans*-silencing.

3.1.1 MARs Do Not Prevent *Trans*-Silencing: Transcriptional Silencing

A simple interpretation of the loop domain model is that MARs protect a transgene from transcriptional gene silencing. However, these predictions need to be modified to explain the results from a series of experiments. As noted above, the original experiments using animal cell lines demonstrated that a transgene flanked by MARs is expressed at levels proportional to transgene copy number. However, Allen et al. (1993) and a subsequent study, which used animal cells (Poljak et al. 1994), demonstrated that lines with many transgene copies were not protected from silencing, even when MARs flanked the transgene.

The finding that MARs, at high copy number, do not prevent gene silencing supports a model in which homology-dependent gene silencing played a major role, especially when the transgene copy number was increased above a "threshold level." Vaucheret et al. (1998) crossed a variety of transgenic plants transformed by MAR-flanked transgenes with V271, a "super-silencer" plant that transcriptionally silences any "target" transgene that is driven by a CaMV 35S promoter (Vaucheret 1993). V271-mediated silencing occurred in *trans*, irrespective of the different target locations or the presence of MARs, suggesting that siRNAs corresponding to the 35S promoter may have been actively recruiting transcriptional silencing machinery to the transgene loci.

Ascenzi et al. (2003) used a similar strategy and crossed lines with additional dominant (or *trans*-based) silencer loci with lines expressing "target" 35S transgene loci at various genomic locations (Ulker et al. 1999). In all of the crosses, the target loci were silenced, regardless of whether or not the Rb7 MAR flanked the target locus. Nuclear run-on assays, which test transcriptional activity, showed that two of the dominant silencer loci, with transgenes lacking MARs (control lines), appeared to be transcriptionally inactive, even though they were able to silence other 35S genes in *trans*. Since siRNAs are needed to maintain heterochromatic areas at centromeres and other regions (Dunoyer et al. 2007; Howell et al. 2007; Kavi et al. 2005), which are not known to be transcribed at detectable levels, a similar mechanism may be in place for the dominant silencing loci. Perhaps during cell division siRNAs are transcribed to help re-establish heterochromatin after DNA replication.

In a separate experiment, Ascenzi et al. crossed the F1 progeny of the dominant silencing line x 35S gene (now silenced) line with a plant expressing P1-HCPro, a viral suppressor of silencing. P1-HCPro is known to reverse RNA-silencing of transgenes (Anandalakshmi et al. 1998 and reviewed in Allen et al. 2000) and is thought to reverse Post-transcriptional Gene Silencing (PTGS) but not Transcriptional Gene Silencing (TGS). In the presence of P1-HCPro, silencing was eliminated in the 35S target genes if they were flanked with MARs. In contrast, silencing could not be reversed in the lines transformed with transgenes lacking the MARs (control lines). These data provide strong evidence that MARs protect against transcriptional gene silencing, perhaps by impeding the formation of heterochromatin. Thus, when MAR-flanked transgenes are silenced, the silencing is likely due to RNAi that targets the transgene mRNA, or a portion of the promoter. In contrast a transgene that lacks MARs is susceptible to a different form of RNAi that is targeted to the transgene promoter (Park et al. 1996) and results in transcriptional silencing. Transgenes that are located in large DNA loops away from the matrix may be more accessible to heterochromatin machinery, and both the gene and the promoter may become silenced. In both cases, silencing requires the presence of RNA that is homologous to either the promoter or the transcribed region. One possibility for the difference in silencing response is that MARs may have prevented RNA polymerase read-through into the promoters of adjacent genes (reviewed in Allen et al. 2000 and in Thompson et al. 2006).

3.1.2   MARs Do Not Prevent *Trans*-Silencing: Post-Transcriptional Silencing

Such data lead to a predication that if MARs prevent a transgene from being transcriptionally silenced, then MAR-flanked transgenes could be expressed at a level that is copy-number dependent if RNAi that is targeted towards the highly expressed transgene mRNA is eliminated. Several *Arabidopsis thaliana* mutants have been found that are altered in the RNAi response (Dalmay et al. 2000; Elmayan et al. 1998). Butaye et al. (2004) transformed mutant *A. thaliana sgs2*[5] and *sgs3*[6] plants (Mourrain et al. 2000) with 35S:*GUS* transgenes either flanked, or not flanked by the chicken lysozyme A element MAR (Phi-Van et al. 1990; Stief et al. 1989). The MAR-flanked transgenes increased GUS expression by fivefold in *sgs2* and 12-fold in *sgs3* plants. Transgene protein levels reached nearly 10% of the total soluble protein in some of the plants.

Butaye et al. (2005) also found that transgene expression in wild-type plants transformed with the MAR-flanked transgene was only 60% of the expression seen in plants transformed with the transgene lacking MARs. While the majority of studies in plants have reported a positive effect on transgene expression when MARs are included in the transgene (reviewed in Allen et al. 2000; and Thompson et al.

---

[5] *sgs2* is *rdr6* or RNA Dependent RNA Polymerase 6.

[6] *sgs3* is a plant- specific protein that is frequently associated with RNA-silencing.

2006), an increase in transgene expression is not always seen. In fact, a decrease in transgene protein or mRNA would be expected if transcription rates drastically increased due to the flanking MARs and RNAi-mediated silencing was triggered. However, when RNAi cannot occur due to a mutation, the full stimulatory effect on expression due to the flanking MARs is revealed (Butaye et al. 2004).

In order for RNAi to occur, a target RNA must be present. However, if a gene is inducible or tissue/cell-specific, the target RNA may not be present at the same time or in the same cell as the silencing signal. If transcriptional activity causes silencing, then transgenes with inducible promoters, such as the tetracycline-inducible 3X promoter (Gatz et al. 1992), may be less susceptible to silencing. In order for RNAi to continue, the process must be actively maintained. To test this hypothesis, Abranches et al. (2005) used an inducible luciferase (*LUC*) reporter transgene with or without flanking Rb7 MARs. The *LUC* transgene was transformed into NT1 tobacco suspension cells using microprojectile bombardment (Allen et al. 1996). The transformed lines were grown either under conditions for continuous transcription of *LUC*, or conditions where *LUC* transcription was initially inactive but induced after approximately 50 cell generations[7] (Gao et al. 1991). Delaying the induction resulted in an initial "burst" in LUC expression in the lines with the MAR-flanked *LUC* transgene, which was then followed by a decrease in expression to levels that were similar to the levels seen in the lines with the continually transcribed 3X:*LUC*. Expression results from the continually expressed transgene were similar to earlier experiments when a 35S:*GUS* transgene was used. Regardless of transcriptional activity, expression of the 3X:*LUC* was higher in lines with MAR-flanking transgenes than in the lines with either a λ DNA-flanked spacer control or the control lacking spacer DNA. These data confirmed that the process of transcription increases silencing, as previously noted (Baulcombe 1996; English and Baulcombe 1997; Que et al. 1997; Vaistij et al. 2002; van Blokland et al. 1997; Vaucheret et al. 1997), especially in the absence of the flanking Rb7 MAR. In addition, the MAR-flanked transgenes could be transcriptionally activated following prolonged inactivity, suggesting that MARs can prevent a transgene from being transcriptionally silenced by the spread of condensed chromatin or from siRNAs produced by readthrough from adjacent genes.

One of the original studies on the use of MARs in plants was by Schoffel et al. (1993), who demonstrated one of the clearest examples of position-independent transgene expression in plants with up to six transgene copies. Schoffl et al. (1993) also used *Agrobacterium* transformation and found a strong effect from a MAR from the soybean heat shock gene *Gmhspl 7.6-L*. With some exceptions, reports of strong MAR effects (~3-fold or greater) on plant transgene expression are less frequent when *Agrobacterium* transformation is used (Annadana et al. 2002; Breyne et al. 1992; Butaye et al. 2004; De Bolle et al. 2003; Halweg et al. 2005; Holmes-Davis and Comai 2002; Kim et al. 2005; Levee et al. 1999; Maximova et al. 2003; Mlynarova et al. 1995, 1994, 2002; Oh et al. 2005; Sidorenko et al. 2003; Van der

---

[7] NT1 generation time is approximately 17 hours.

Geest et al. 1994; Van Leeuwen et al. 2001). In principle, use of induction following several rounds of cell division is similar to the approach used by Abranches et al. (2005). Schoffl et al. (1993) excised tobacco leaves, which were treated with a 2-h heat shock at 40°C. In contrast to genes that are constantly being transcribed and therefore more susceptible to RNAi, the use of induction results in a transgene being transcribed in the absence of RNAi targeted towards the transgene. The flanking MARs act to prevent the spread of silencing through the transgene, thus keeping it poised for transcription following induction.

3.1.3 MARs Prevent *Cis*-Silencing

The model presented for how MARs improve transgene expression can best be described as minimizing gene silencing (reviewed by Allen et al. 2000). Until recently, the only data available for how MARs function has been circumstantial. For example, silencing can be reversed in lines that contain MARs flanking their transgene when crossed with plants expressing the viral suppressor of RNA VIGS, P1-HCPro (Anandalakshmi et al. 1998). In addition, MAR-flanked transgenes are expressed at a higher level than the same transgenes lacking MARs when transformed into *A. thaliana* with mutations in some of the key components of the RNAi pathway for sense-transgene silencing (Butaye et al. 2004). Additional data that support the model include findings that MARs do not protect against *trans*-silencing (Vaucheret et al. 1998) and that MAR-flanked transgenes are silenced when the transgene copy number reaches a threshold value (Allen et al. 1996; Allen et al. 1993). While these data are suggestive, they still only correlate MARs with protection against transgene silencing.

Conclusive data would require being able to examine a transgene with and without MARs at a defined position within the genome to avoid the possible variation due to differences in the chromatin structure at the site of integration. Regardless of how many transgenic lines are tested, it has not been possible to test MARs at specific locations in the genome due largely to the extremely low efficiency of gene targeting in higher plants (Britt and May 2003; Hanin and Paszkowski 2003; Kumar et al. 2006). However, it is possible to excise DNA with the use of site-specific recombination. Fiering et al. (1993) used the site-specific Flp recombinase that can excise DNA when the Flp Target Recognition (FRT) sites flank the DNA in direct repeats. This strategy allowed Fiering et al. (1993) to compare the effect of various regulatory regions from the β-globin Locus Control Region on the expression of a transgene. Initially, the transgene and regulatory region flanked by FRT was transformed into the genome. The transgene expression levels and other characteristics, such as DNase I sensitivity, can then be tested. The Flp recombinase is then expressed in the cell by transient expression, to excise the particular regulatory element being tested. Transgene expression and the other characteristics can then again be analyzed at the same location. The difference in expression can then be directly attributed to the presence or absence of the regulatory element.

Mlynarova et al. (2003) transformed tobacco with a 35S:*GUS* reporter gene with the chicken lysozyme A element MAR. The MARs flanked both the 35S:*GUS* reporter and the selection gene, nos promoter:*nptII*. One flanking MAR was flanked by the wild-type *loxP* and the second MAR was flanked by a mutated form of *loxP* (*lox511*). In principle the wild-type *loxP* can only recombine with wild-type *loxP* and the *lox511* can only recombine with *lox511*. In the presence of Cre recombinase, which had been introduced by re-transformation with 35S:*Cre*, either *loxP* or *lox511* self-recombination would result in the excision of the specific MAR flanking each transgene, *GUS* and *nptII*. It was anticipated that the resulting progeny would be missing a MAR from either flank of the reporter gene-selection gene domain. In many of the progeny the mutated *lox511* and wild-type *loxP* had recombined, however, it was possible to find one progeny line was missing a single MAR, line AGCNA-61 (see Fig. 3).

A comparison of the AGCNA-61 *GUS* expression showed that when the 3′ MAR was present *GUS* expression was retained throughout the life of the plant. However, AGCNA-61 plants in which the MAR had been excised showed several characteristics that are typically found in plants silenced by RNAi. For example, a comparison of *GUS* expression at different times showed that after 3 months the AGCNA-61 plants lacking the MAR showed a strong loss in GUS activity, and the silencing was intensified in homozygous plants. Nuclear run-on assays confirmed that the 35S:*GUS* transgene in AGCNA-61 plants remained transcriptionally active even when

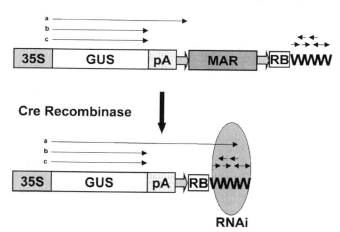

Fig. 3 MARs prevent readthrough transcription and *Cis* silencing. (*Top*) The 35S:*GUS* transgene with a nos polyadenylation signal (pA). Transcripts are produced as shown by *arrows* **a**, **b**, and **c**. The A element chicken lysozyme MAR at the 3′ end of the transgene cassette blocks readthrough transcript **a** from continuing into the highly condensed plant DNA (**W**) flanking the transgene. The mutant *loxP511* sites (*wide arrows*) are organized as direct repeats to flank the MAR. (*Bottom*) In the presence of Cre recombinase the directly repeated *loxP511* sites recombine and the MAR is excised from the genome. Transcript **a** readthrough continues into the highly condensed plant flanking DNA. Small RNAs (*small arrows*) produced within the flanking DNA form double-stranded RNA with homology to readthrough the 3′ end of transcript **a** and RNAi (*grey oval*) is initiated, which spreads into and silences the 35S:*GUS* transgene

the MAR was absent. When the same plants were tested by Real Time-Polymerase Chain Reaction (RT-PCR), the AGCNA-61 plants lacking the MAR had a greatly reduced level of *GUS* mRNA. Importantly, when the same plants were compared for the presence of siRNAs for *GUS*, only the AGCNA-61 plants lacking the MAR showed 21–24-bp RNAs. The presence of the 21–24-bp RNAs is found in plants that have active RNAi (Hamilton and Baulcombe 1999; Hamilton et al. 2002), which in this case was targeted against the *GUS* mRNA. This finding supports the hypothesis that MARs protect transgenes from *cis*-silencing by RNAi. As noted earlier, RNAi has been shown to be key for the initiation and maintenance of heterochromatin.

3.1.4 *Agrobacterium* Transformation and Selection Bias

As noted in the previous section, one reason why MARs appear to improve transgene expression from some gene cassettes used for transformation but have no effect in others may be explained by the different types of gene silencing that MARs prevent. However, differences in the method of transformation may also play an important role. The greatest increases in transgene expression seen in experiments have resulted from transformants produced by direct DNA transformation, such as microprojectile bombardment. It is generally thought that the use of direct DNA transformation results in a higher transgene copy number than when *Agrobacterium* transformation is used. It is well known that higher transgene copy numbers have an increased probability of being silenced, or have greater expression variation (Hobbs et al. 1990, 1993; Koprek et al. 2001).

The use of direct DNA transformation does have an important advantage for studying the effect of MARs on transgene expression because the selectable marker transgene and the reporter transgene can be physically separated and a co-transformation procedure can be used in which the separate marker and the reporter genes are mixed prior to transformation. By unlinking the selection and the reporter genes, the transformation process for each is independent, and there is no selection bias for strong or against weak gene expression. In practice, transformants are first identified by resistance to the selective agent and then are measured for reporter gene activity. By physically unlinking the selection and the reporter gene, integration of the reporter gene can occur into different genomic locations and still be recovered so that possible differences in transgene copy number are meaningful.

While co-transformation can also be done using *Agrobacterium* (Komari et al. 1996; De Neve et al. 1997; Iradani 1998; Vain et al. 2003), all of the MAR studies to date have been done using binary vectors with the selection and reporter transgenes physically linked (reviewed in Allen et al. 2000; and Thompson et al. 2006). A typical *Agrobacterium* transformation involves infection of the competent plant tissue, killing of the *Agrobacterium*, and selecting for cells that can survive the antibiotic or herbicide due to the presence of the selection transgene. It is critical that the selectable marker transgene be expressed in order for the cells to survive and proliferate. Cells that are "transformed" but do not express the selectable marker are routinely discarded. The use of selection therefore creates a bias by selecting only for transformants with active transgenes which must be integrated into regions of the genome that are transcriptionally active at the time of selection, such as euchromatin. Of course, the linked reporter transgene integrates into the same active genomic region and thus the integration sites cannot be random.

According to the loop domain model, the importance of flanking MARs becomes most evident when a transgene integrates into inactive, condensed regions of the genome, such as heterochromatin. If a method of identifying transformed cells that does not rely on selection is used it then becomes possible to achieve more random integration. A recent study by Francis and Spiker (2005) demonstrated how selection bias may impact our interpretation by using a PCR method to identify transgenic plants with integrated *nptII* selection and *GUS* reporter genes. When transformants that had been identified by PCR were tested for kanamycin resistance, which requires *nptII* expression, approximately 30% were sensitive indicating that the selection and linked reporter gene had integrated into a region of the genome that was prone to gene silencing. These results were confirmed recently by Kim et al. (2007) who transformed A. *thaliana* Col-0 suspension cells in the absence of selection. Transformants were identified with T-DNA integration events that had a higher probability of occurring in heterochromatic regions, such as centromeres, telomeres, and rDNA repeat regions.

3.1.5   Do MARs Impact Transgene Integration Events?

Selection requires active expression of the selection gene. When selection is used, transgene integration is highly likely to occur in transcriptionally poised or active genomic regions. If MARs are associated with regions of active transcription, would a higher percentage of the selection-derived transformants integrate in close proximity to endogenous MARs, especially when the transgene construct used lacks flanking MARs? Second, if transformants could be identified without the use of selection, would such transgenes be found in genomic regions lacking local endogenous MARs? Of course, for truly random integration to occur, it would be necessary to assume that the endogenous MARs do not enhance integration site preference. Interestingly, several studies have found either predicted (Makarevitch et al. 2003), or experimentally tested MARs (Dietz et al. 1994; Iglesias et al. 1997; Sawasaki et al. 1998; Takano et al. 1997) in close proximity to the site of transgene integration, but selection has typically been used to identify these transformants. The interpretation from such results is that transgenes integrate near endogenous MARs, but to truly address these questions, it is necessary to study integration sites from transformants identified without selection bias.

Koprek et al. (2001) described a novel approach that may provide some of the answers to this question. The *Agrobacterium* transformation included a selection step to identify the primary transformants with the transgene integrated at the "launchpad" integration site. However, an AC/DS transposon system was then used to generate novel genomic integration sites, resulting from transposition, without using selection. This procedure allows the DS element, carrying a Bar gene (*Basta* *r*esistance), to randomly integrate into new genomic positions during the transposition process. Initially the Bar transgene, flanked by inverted-repeat DS ends (Bar-DS), was transformed into barley using *Agrobacterium* and transformants were selected on the herbicide Basta. The Basta-resistant transformants were then crossed with a plant expressing the AC transposase gene and the F_1 progeny were then self-pollinated. The F_2 plants were then re-tested for Basta resistance. When transposition occurred, 25% of the new integration sites were unlinked from the original site, indicating that *Bar* had integrated into different chromosomes.

A subsequent study by Zhao et al. (2006) mapped the initial Bar-DS integration sites and the new integration sites following transposition. Sequence analysis of the initial T-DNA integration sites, which were selected for by Basta, found a sevenfold greater preference for integration into non-redundant, gene-rich regions of the barley genome versus non-gene regions. In contrast, when the new sites resulting from DS elements were mapped, a preference for regions containing endogenous MARs was found. It is possible that the DS transposon prefers more accessible chromatin with MARs, or that regions containing MARs are generally favored sites for the integration of new DNA. This could occur, for example, if DNA repair enzymes were more likely to be found in active parts of the genome.

Petersen et al. (2002) transformed barley with 35S:*GUS* transgenes flanked by either the P1 MAR from soybean (Breyne et al. 1992) or the Transformation Booster Sequence (TBS) MAR from petunia (Buising and Benbow 1994; Galliano et al. 1995; Meyer et al. 1988). In vitro MAR binding assays using nuclear matrix isolated from barley leaves showed that the P1 MAR bound strongly whereas the TBS MAR binding was insignificant. Interestingly, both MARs increased the transformation efficiency by approximately twofold, but only the P1 MAR increased expression (12- to 13-fold). The mechanism for how both the P1MAR and the TBS MAR stimulate transformation is unknown but not completely unexpected. Because, as Zhao et al. (2006) noted, though DS elements integrate preferentially near endogenous MARs, the converse may also be true: MARs that are associated with transgenes may facilitate integration.

4   The Use of MARs to Improve the Stability of Transgene Expression

4.1   MARs for Protein Overexpression

Much of the research on MARs has focused on their practical use as tools for preventing the *cis*-silencing of transgenes. MARs can be used in combination with other strategies such as plants with mutations in silencing components (Butaye et al. 2004) or plants transformed with viral suppressors of silencing (Anandalakshmi et al. 1998; Marathe et al. 2000) that block, in *trans*, RNA silencing and achieve high levels of transgene expression. Several examples have been published that have used MARs in combination with a gene of interest for practical applications such as reducing photorespiration to improve photosynthesis and biomass production (Kebeish et al. 2007; Khan 2007; Leegood 2007; Niessen et al. 2007), increasing expression of a starch-branching enzyme (Kim et al. 2005), or increasing stilbene synthase expression (Jin et al. 2004). While MARs can be used for increasing the level of expression of a useful gene product, very high expression levels may also show an increased likelihood of initiating RNA silencing. It is highly likely that reports from various labs in which MARs have little or even negative effects in certain transformation experiments are due to post-transcriptional gene silencing (Baulcombe 1999).

4.2   MARs for Stabilizing RNAi

Ironically, according to the model, including MARs in a transgene construct may be one of the best ways to stabilize *trans*-silencing through post-transcriptional gene silencing, or RNAi. Gene knockdown through RNAi is increasingly becoming important as a method for understanding gene function (McGinnis et al. 2007; Waterhouse and Helliwell 2003) and for metabolic pathway engineering (Chen et al. 2005; Kotting et al. 2005; Kusaba 2004; Tang and Galili 2004). Many recent applications use RNAi constructs to knockdown the expression of an endogenous gene through the use of hairpin constructs (Waterhouse and Helliwell 2003). In order for RNAi to continue to function it is necessary for the hairpin construct to continue to be transcriptionally active. In some cases the RNAi vector becomes transcriptionally silenced, leading to loss of the knockdown phenotype (Fojtova et al. 2003; Kerschen et al. 2004). MARs work well for preventing *cis*-silencing of the promoter and have been shown to stabilize the transcriptional activity of cassettes containing viral genes for silencing incoming viral sequences vectors for producing virus-resistant plants (Robertson 2004).

One specific example of the benefit of using MARs to stabilize silencing comes from work in tobacco, testing resistance to the tospovirus Tomato Spotted Wilt Virus (TSWV) (Levin et al. 2005). A transgene cassette designed for constitutive expression of the TSWV 35S: nucleocapsid protein gene, TSWV-N with and without flanking Rb7 MARs was used to transform tobacco. When a total of 79 MAR and 66 non-MAR events were tested in the field over four generations, the MAR-containing plants showed significantly higher levels of resistance to TSWV, which is transmitted throughout the field season by thrips. These findings validate the prediction that PTGS plays an active role in the final expression levels of MAR-flanked genes and that before concluding that MARs themselves have a negative effect on gene expression, silencing must be tested or eliminated as a variable. Levin et al. (2005) showed that resistant lines produced small RNAs of 21–25 bp and that they maintained transcriptional activity as demonstrated by nuclear run-on assays. The Levin et al. experiment clearly demonstrates the value of MARs for preserving transcriptional activity and maintaining virus resistance.

5 Conclusion

Much of our current understanding of the role MARs play in higher-order chromatin structure and nuclear architecture comes from research with yeast and animal cells, while the research on MARs in plants has been primarily limited to whether including MARs in transgene constructs can be used as a tool to stabilize transgene expression. While understanding whether MARs improve transgene expression is important, MARs are also thought to be important for other basic nuclear processes such as DNA replication and nuclear organization. New research shows that the non-coding regions of the genome that were previously called "junk DNA" are critical for controlling the expression of genes.

Genome sequencing has enabled researchers to begin to see how genes and non-coding regions are organized and to begin to make predictions based on similarity. However, much work remains to be done in order to understand how the sequences are arranged in a three-dimensional space. A computer algorithm has been developed to predict MARs (Rudd et al. 2004; Tetko et al. 2006) and it has been used to predict the MARs in the *Arabidopsis thaliana* genome. However, when the algorithm was used to predict known plant MARs only 60% were "found." In contrast, an earlier algorithm, called MAR-finder (Singh et al. 1997) was able to predict nearly 80% of the known plant MARs. A major limitation for both algorithms is the lack of experimental data. New genomic approaches that allow the identification and mapping of large numbers of sequences that bind to the matrix may eventually provide such data (Linnemann et al. 2007).

Finally, the field of nuclear organization and the nuclear matrix remains controversial and much of the controversy is fueled by our lack of knowledge about the nuclear matrix and whether it even exists as a discrete entity (Pederson 2000). New tools are desperately needed to compare the properties of the nuclear matrix in different tissues, over time, and as the nucleus changes shape and/or undergoes endoreduplication. What is the role of nuclear actin? What role does RNA play in the nuclear matrix? How are MARs recognized, and what is the biological significance of MARs with different binding affinities? New methods will be needed to examine large-scale protein–protein interactions in living cells, and to begin to unravel the role of RNA both as a catalyst and as a structural component (Rozowsky et al. 2007; Washietl et al. 2007). With the current rate of advances in technology and information processing, the future for understanding how basic units of genetic information are maintained in three-dimensional space, over time, and how their structure relates to function, is very promising.

References

Abranches R, Shultz RW, Thompson WF, Allen GC (2005) Matrix attachment regions and regulated transcription increase and stabilize transgene expression. Plant Biotech J 3:535–543

Aguilera A (2005) Cotranscriptional mRNP assembly: from the DNA to the nuclear pore. Curr Opin in Cell Biol 17:242–250

Alberts B, Johnson A, Lewis J, Raff M, Roberts K, Walter P (2002) Molecular biology of the cell. Garland, New York

Alberts B, Sternglanz R (1990) Chromatin contract to silence. Nature 344:193–194

Allen GC, Hall G, Michalowski S, Newman W, Spiker S, Weissinger AK, Thompson WF (1996) High-level transgene expression in plant cells: effects of a strong scaffold attachment region from tobacco. Plant Cell 8:899–913

Allen GC, Hall GE Jr, Childs LC, Weissinger AK, Spiker S, Thompson WF (1993) Scaffold attachment regions increase reporter gene expression in stably transformed plant cells. Plant Cell 5:605–613

Allen GC, Spiker S, Thompson WF (2000) Use of matrix attachment regions (MARs) to minimize transgene silencing. Plant Mol Biol 43:361–376

Allen ND, Cran DG, Barton SC, Hettle S, Reik W, Surani MA (1988) Transgenes as probes for active chromosomal domains in mouse development. Nature 333:852–855

Amati BB, Gasser SM (1988) Chromosomal ARS and CEN elements bind specifically to yeast nuclear scaffold. Cell 54:967–978

Anandalakshmi R, Pruss GJ, Ge X, Marathe R, Mallory AC, Smith TH, Vance VB (1998) A viral suppressor of gene silencing in plants. Proc Natl Acad Sci USA 95:13079–13084

Annadana S, Mlynarova L, Udayakumar M, de Jong J, Nap JP (2002) The potato Lhca3.St.1 promoter confers high and stable transgene expression in chrysanthemum, in contrast to CaMV-based promoters. Mol Breed 8:335–344

Ascenzi R, Ulker B, Todd JJ, Sowinski DA, Schimeneck CR, Allen GC, Weissinger AK, Thompson WF (2003) Analysis of trans-silencing interactions using transcriptional silencers of varying strength and targets with and without flanking nuclear matrix attachment regions. Trans Res 12:305–318

Assaad FF, Tucker KL, Signer ER (1993) Epigenetic repeat-induced silencing (RIGS) in Arabidopsis. Plant Mol Biol 22:1067–1085

Baulcombe DC (1996) RNA as a target and an initiator of post-transcriptional gene silencing in transgenic plants. Plant Mol Biol 32:79–88

Baulcombe DC (1999) Gene silencing: RNA makes RNA makes no protein. *Curr Biol* 9:R599–R601

Berezney R, Coffey DS (1974) Identification of a nuclear protein matrix. Biochem Biophys Res Commun 60:1410–1417

Bi X, Yu Q, Sandmeier JJ, Zou Y (2004) Formation of boundaries of transcriptionally silent chromatin by nucleosome-excluding structures. Mol Cell Biol 24:2118–2131

Bode J, Maass K (1988) Chromatin domain surrounding the human interferon-b gene as defined by scaffold-attached regions. Biochem 27:4706–4711

Bodnar JW (1988) A domain model for eukaryotic DNA organization: a molecular basis for cell differentiation and chromosome evolution. J Theor Biol 132:479–507

Bonifer C, Hecht A, Saueressig H, Winter DM, Sippel AE (1991) Dynamic chromatin: the regulatory domain organization of eukaryotic gene loci. J Cell Biochem 47:99–108

Bourdon V, Ladbrooke Z, Wickham A, Lonsdale D, Harwood W (2002) Homozygous transgenic wheat plants with increased luciferase activity do not maintain their high level of expression in the next generation. Plant Sci 163:297–305

Bregitzer P, Tonks D (2003) Inheritance and expression of transgenes in barley. Crop Sci 43:4–12

Breyne P, van Montagu M, Depicker A, Gheysen G (1992) Characterization of a plant scaffold attachment region in a DNA fragment that normalizes transgene expression in tobacco. Plant Cell 4:463–471

Brickner DG, Cajigas I, Fondufe-Mittendorf Y, Ahmed S, Lee P-C, WidomJ , Brickner JH (2007) H2A.Z-mediated localization of genes at the nuclear periphery confers epigenetic memory of previous transcriptional state. PLoS Biol 5:e81

Britt AB, May GD (2003) Re-engineering plant gene targeting. Trends Plant Sci 8:90–95

Buising CM, Benbow RM (1994) Molecular analysis of transgenic plants generated by micropro-jectile bombardment – effect of petunia transformation booster sequence. Mol Gen Genet 243:71–81

Butaye KMJ, Goderis IJWM, Wouters PFJ, Pues JM-TG, Delaure SL, Broekaert WF, Depicker A, Cammue BPA, De Bolle MFC (2004) Stable high-level transgene expression in *Arabidopsis thaliana* using gene silencing mutants and matrix attachment regions. Plant J 39:440–449

Butaye KMJ, Cammue BPA, Delaure, SL, De Bolle MFC (2005) Approaches to minimize varia-tion of transgene expression in plants. Mol Breeding 16:79–81

Cabal GG, Genovesio A, Rodriguez-Navarro S, Zimmer C, Gadal O, Lesne A, Buc H, Feuerbach-Fournier F, Olivo-Marin J-C, Hurt EC, Nehrbass U (2006) SAGA interacting factors confine sub-diffusion of transcribed genes to the nuclear envelope. Nature 441:770–773

Callard D, Mazzolini L (1997) Identification of proliferation-induced genes in *Arabidopsis thal-iana*. Characterization of a new member of the highly evolutionarily conserved histone H2A.F/Z variant subfamily. Plant Physiol 115:1385–1395

Casolari JM, Brown CR, Komili S, West J, Hieronymus H, Silver PA (2004) Genome-wide locali-zation of the nuclear transport machinery couples transcriptional status and nuclear organiza-tion. Cell 117:427–439

Chareonpornwattana S, Thara KV, Wang L, Datta SK, Panbangred W, Muthukrishnan S (1999) Inheritance, expression, and silencing of a chitinase transgene in rice. Theor Appl Genet 98:371–378

Chen S, Hajirezaei M, Peisker M, Tschiersch H, Sonnewald U, Bornke F (2005) Decreased sucrose-6-phosphate phosphatase level in transgenic tobacco inhibits photosynthesis, alters carbohydrate partitioning, and reduces growth. Planta 221:479–492

Choi K, Park C, Lee J, Oh M, Noh B, Lee I (2007) Arabidopsis homologs of components of the SWR1 complex regulate flowering and plant development. Develop 134:1931–1941

Chua YL, Brown AP, Gray JC (2001) Targeted histone acetylation and altered nuclease accessibil-ity over short regions of the pea plastocyanin gene. Plant Cell 13:599–612

Chua YL, Watson LA, Gray JC (2003) The transcriptional enhancer of the pea plastocyanin gene associates with the nuclear matrix and regulates gene expression through histone acetylation. Plant Cell 15:1468–1479

Cockerill PN, Garrard WT (1986) Chromosomal loop anchorage of the kappa immunoglobulin gene occurs next to the enhancer in a region containing topoisomerase II sites. Cell 44:273–282

Cole CN, Scarcelli JJ (2006) Transport of messenger RNA from the nucleus to the cytoplasm. Curr Opin Cell Biol 18:299–306

Cook PR (1989) The nucleoskeleton and the topology of transcription. Eur J Biochem 185:487–501

Crawford GE, Davis S, Scacheri PC, Renaud G, Halawi MJ, Erdos MR, Green R, Meltzer PS, Wolfsberg TG, Collins FS (2006) DNase-chip: a high-resolution method to identify DNase I hypersensitive sites using tiled microarrays. Nat Methods 3:503–509

Cuvier O, Hart CM, Kas E, Laemmli UK (2002) Identification of a multicopy chromatin boundary element at the borders of silenced chromosomal domains. Chromosoma 110:519–531

Dalmay T, Hamilton A, Rudd S, Angell S, Baulcombe DC (2000) An RNA-dependent RNA polymerase gene in Arabidopsis is required for posttranscriptional gene silencing mediated by a transgene but not by a virus. Cell 101:543–553

De Bolle MFC, Butaye KMJ, Coucke WJW, Goderis I, Wouters PFJ, van Boxel N, Broekaert WF, Cammue BPA (2003) Analysis of the influence of promoter elements and a matrix attachment region on the inter-individual variation of transgene expression in populations of *Arabidopsis thaliana*. Plant Sci 165:169–179

De Neve M, De Buck S, De Wilde C, Van Houdt H, Strobbe I, Jacobs A, Van M, Depicker A (1997) T-DNA integration patterns in co-transformed plant cells suggest that T-DNA repeats originate from co-integration of separate T-DNAs. Plant J 11:15–29

De Neve M, De Buck S, De Wilde C, Van Houdt H, Strobbe I, Jacobs A, Van M, Depicker A (1999) Gene silencing results in instability of antibody production in transgenic plants. Mol Gen Genet 260:582–592

Deal RB, Kandasamy MK, McKinney EC, Meagher RB (2005) The nuclear actin-related protein ARP6 is a pleiotropic developmental regulator required for the maintenance of FLOWERING LOCUS C expression and repression of flowering in Arabidopsis. Plant Cell 17:2633–2646

Deal RB, Topp CN, McKinney EC, Meagher RB (2007) Repression of flowering in Arabidopsis requires activation of FLOWERING LOCUS C expression by the histone variant H2A.Z. Plant Cell 19:74–83

Dean C, Jones J, Favreau M, Dunsmuir P, Bedbrook J (1988) Influence of flanking sequences on variability in expression levels of an introduced gene in transgenic tobacco plants. Nucleic Acids Res 16:9267–9283

Dietz A, Kay V, Schlake T, Landsmann J, Bode J (1994) A plant scaffold attacked region detected close to a T-DNA integration site is active in mammalian cells. Nucleic Acids Res 22:2744–2751

Dilworth DJ, Tackett AJ, Rogers RS, Yi EC, Christmas RH, Smith JJ, Siegel AF, Chait BT, Wozniak RW, Aitchison JD (2005) The mobile nucleoporin Nup2p and chromatin-bound Prp20p function in endogenous NPC-mediated transcriptional control. J Cell Biol 171:955–965

Dorer DR, Henikoff S (1997) Transgene repeat arrays interact with distant heterochromatin and cause silencing in cis and trans. Genetics 147:1181–1190

Dunoyer P, Himber C, Ruiz-Ferrer V, Alioua A, Voinnet O (2007) Intra- and intercellular RNA interference in *Arabidopsis thaliana* requires components of the microRNA and heterochromatic silencing pathways. Nat Genet 39:848–856

Elmayan T, Balzergue S, Beon F, Bourdon V, Daubremet J, Guenet Y, Mourrain P, Palauqui JC, Vernhettes S, Vialle T, Wostrikoff K, Vaucheret H (1998) Arabidopsis mutants impaired in cosuppression. Plant Cell 10:1747–1757

English JJ, Baulcombe DC (1997) The influence of small changes in transgene transcription on homology-dependent virus resistance and gene silencing. Plant J 12:1311–1318

Fiering S, Kim C, Epner E, Groudine M (1993) An "in–out" strategy using gene targeting and FLP recombinase for the functional dissection of complex DNA regulatory elements: analysis of the β-globin locus control region. Proc Natl Acad Sci USA 90:8469–8473

Fojtova M, Van Houdt H, Depicker A, Kovarik A (2003) Epigenetic switch from posttranscriptional to transcriptional silencing is correlated with promoter hypermethylation. Plant Physiol 133:1240–1250

Francis KE, Spiker S (2005) Identification of *Arabidopsis thaliana* transformants without selection reveals a high occurrence of silenced T-DNA integrations. Plant J 41:464–477

Galliano H, Muller AE, Lucht JM, Meyer P (1995) The transformation boaster sequence from *Petunia hybrida* is a retrotransposon derivative that binds to the nuclear scaffold. Mol Gen Genet 247:614–622

Gao J, Lee JM, An G (1991) The stability of foreign protein production in genetically modified plant cells. Plant Cell Rep 10:533–536

Gasser SM, Laemmli UK (1986) The organization of chromatin loops: characterization of a scaffold attachment site. EMBO J 5:511–518

Gaszner M, Felsenfeld G (2006) Insulators: exploiting transcriptional and epigenetic mechanisms. Nat Rev Genet 7:703–713

Gatz C, Claus F, Regina W (1992) Stringent repression and homogeneous de-repression by tetracycline of a modified CaMV 35S promoter in intact transgenic tobacco plants. Plant J 2:397–404

Gendrel AV, Lippman Z, Martienssen R, Colot V (2005) Profiling histone modification patterns in plants using genomic tiling microarrays. Nat Methods 2:213–218

Gendrel AV, Lippman Z, Yordan C, Colot V, Martienssen RA (2002) Dependence of heterochromatic histone H3 methylation patterns on the Arabidopsis gene *DDM1*. Science 297:1871–1873

Gepts P (2002) A comparison between crop domestication, classical plant breeding, and genetic engineering. Crop Sci 42:1780–1790

Giresi PG, Lieb JD (2006) How to find an opening (or lots of them). Nat Methods 3:501–502

Goetze S, Baer A, Winkelmann S, Nehlsen K, Seibler J, Maass K, Bode J (2005) Performance of genomic bordering elements at predefined genomic loci. Mol Cell Biol 25:2260–2272

Greally JM (2007) Genomics: encyclopaedia of humble DNA. Nature 447:782–783

Greb T, Mylne JS, Crevillen P, Geraldo N, An H, Gendall AR, Dean C (2007) The PHD finger protein VRN5 functions in the epigenetic silencing of Arabidopsis FLC. Curr Biol 17:73–78

Gromova II, Nielsen OF, Razin SV (1995) Long-range fragmentation of the eukaryotic genome by exogenous and endogenous nucleases proceeds in a specific fashion via preferential DNA cleavage at matrix attachment sites. J Biol Chem 270:18685–18690

Grosveld F, van Assendelft GB, Greaves DR, Kollias G (1987) Position-independent, high-level expression of the human b-globin gene in transgenic mice. Cell 51:975–985

Hall GE, Jr., Allen GC, Loer DS, Thompson WF, Spiker S (1991) Nuclear scaffolds and scaffold-attachment regions in higher plants. Proc Natl Acad Sci USA 88:9320–9324

Hall IM, Shankaranarayana GD, Noma K, Ayoub N, Cohen A, Grewal SI (2002) Establishment and maintenance of a heterochromatin domain. Science 297:2232–2237

Halweg C, Thompson WF, Spiker S (2005) The Rb7 matrix attachment region increases the likelihood and magnitude of transgene expression in tobacco cells: a flow cytometric study. Plant Cell 17:418–429

Hamilton A, Voinnet O, Chappell L, Baulcombe D (2002) Two classes of short interfering RNA in RNA silencing. EMBO J 21:4671–4679

Hamilton AJ, Baulcombe DC (1999) A species of small antisense RNA in post transcriptional gene silencing in plants. Science 286:950–952

Hanin M, Paszkowski J (2003) Plant genome modification by homologous recombination. Curr Opin Plant Biol 6:157–162

Henderson IR, Jacobsen SE (2007) Epigenetic inheritance in plants. Nature 447:418–424

Heng HH, Goetze S, Ye CJ, Liu G, Stevens JB, Bremer SW, Wykes SM, Bode J, Krawetz SA (2004) Chromatin loops are selectively anchored using scaffold/matrix-attachment regions. J Cell Sci 117:999–1008

Henikoff S (1998) Conspiracy of silence among repeated transgenes. Bioessays 20:532–535

Hobbs SLA, Kpodar P, Delong CMO (1990) The effect of T-DNA copy number, position and methylation on reporter gene expression in tobacco transformants. Plant Mol Biol 15:851–864

Hobbs SLA, Warkentin TD, DeLong CMO (1993) Transgene copy number can be positively or negatively associated with transgene expression. Plant Mol Biol 21:17–26

Holmes-Davis R, Comai L (2002) The matrix attachment regions (MARs) associated with the heat shock cognate 80 gene (HSC80) of tomato represent specific regulatory elements. Mol Gen Gen 266:891–898

Howell MD, Fahlgren N, Chapman EJ, Cumbie JS, Sullivan CM, Givan SA, Kasschau KD, Carrington JC (2007) Genome-wide analysis of the RNA-DEPENDENT RNA POLYMERASE6/DICER-LIKE4 pathway in Arabidopsis reveals dependency on miRNA- and tasiRNA-directed targeting. Plant Cell 19:926–942

Iglesias VA, Moscone EA, Papp I, Neuhuber F, Michalowski S, Phelan T, Spiker S, Matzke M, Matzke AJM (1997) Molecular and cytogenetic analyses of stably and unstably expressed transgene loci in tobacco. Plant Cell 9:1251–1264

Iradani T, Bogani P, Mengoni A, Mastromei G, Buiatti M (1998) Construction of a new vector conferring methotrexate resistance in Nicotiana tabacum plants. Plant Mol Biol 37:1079–1084

Ishii K, Arib G, Lin C, Van Houwe G, Laemmli UK (2002) Chromatin boundaries in budding yeast: the nuclear pore connection. Cell 109:551–562

Istomina NE, Shushanov SS, Springhetti EM, Karpov VL, Krasheninnikov IA, Stevens K, Zaret KS, Singh PB, Grigoryev SA (2003) Insulation of the chicken b-globin chromosomal domain from a chromatin-condensing protein, MENT. Mol Cell Biol 23:6455–6468

Jacobsen SE (1999) Gene silencing: maintaining methylation patterns. Curr Biol 9:R617–R619

James VA, Worland B, Snape JW, Vain P (2004) Strategies for precise quantification of transgene expression levels over several generations in rice. J Exp Bot 55:1307–1313

Jin C, Felsenfeld G (2007) Nucleosome stability mediated by histone variants H3.3 and H2A.Z. Genes Dev 21:1519–1529

Jin Z, Shu-Jun L, Si-Song M, Wei Y, Yuan-Lei H, Qiang W, Zhong-Ping L (2004) Effect of Matrix Attachment Regions on resveratrol production in tobacco with transgene of stilbene synthase from *Parthenocissus henryana*. Acta Bot Sin 46:948–954

Johnson L, Cao X, Jacobsen S (2002) Interplay between two epigenetic marks. DNA methylation and histone H3 lysine 9 methylation. Curr Biol 12:1360–1367

Jorgensen RA, Cluster PD, English J, Que QD, Napoli CA (1996) Chalcone synthase cosuppression phenotypes in petunia flowers: comparison of sense vs. antisense constructs and single-copy vs. complex T-DNA sequences. Plant Mol Biol 31:957–973

Kamakaka RT, Biggins S (2005) Histone variants: deviants? Genes Dev 19:295–310

Kavi HH, Xie W, Fernandez HR, Birchler JA (2005) Global analysis of siRNA-mediated transcriptional gene silencing. Bioessays 27:1209–1212

Kebeish R, Niessen M, Thiruveedhi K, Bari R, Hirsch H-J, Rosenkranz R, Stabler N, Schonfeld B, Kreuzaler F, Peterhansel C (2007) Chloroplastic photorespiratory bypass increases photosynthesis and biomass production in *Arabidopsis thaliana*. Nat Biotech 25:593–599

Kerschen A, Napoli CA, Jorgensen RA, Muller AE (2004) Effectiveness of RNA interference in transgenic plants. FEBS Lett 566:223–228

Khan MS (2007) Engineering photorespiration in chloroplasts: a novel strategy for increasing biomass production. Trends Biotech 25:437–440

Kim S-I, Veena, Gelvin SB (2007) Genome-wide analysis of Agrobacterium T-DNA integration sites in the Arabidopsis genome generated under non-selective conditions. Plant J 51:779–791

Kim WS, Kim J, Krishnan HB, Nahm BH (2005) Expression of *Escherichia coli* branching enzyme in caryopses of transgenic rice results in amylopectin with an increased degree of branching. Planta 220:689–695

Kodama Y, Nagaya S, Shinmyo A, Kato K (2007) Mapping and characterization of DNase I hypersensitive sites in Arabidopsis chromatin. Plant Cell Physiol 48:459–470

Komari T, Hiei T, Saito Y, Murai N, Kumashiro T (1996) Vectors carrying two separate T-DNAs for co-transformation of higher plants mediated by *Agrobacterium tumefaciens* and segregation of transformants free from selection markers. Plant J 10:165–174

Koprek T, Rangel S, McElroy D, Louwerse JD, Williams-Carrier RE, Lemaux PG (2001) Transposon-mediated single-copy gene delivery leads to increased transgene expression stability in barley. Plant Physiol 125:1354–1362

Kotting O, Pusch K, Tiessen A, Geigenberger P, Steup M, Ritte G (2005) Identification of a novel enzyme required for starch metabolism in Arabidopsis leaves. The phosphoglucan, water dikinase. Plant Physiol 137:242–252

Kumar S, Franco M, Allen GC (2006) Gene targeting: development of novel systems for genome engineering in plants. In: Teixeira da Silva JA (ed) Floriculture, ornamental and plant biotechnology: advances and topical issues. Global Science Books, London, pp 84–98

Kusaba M (2004) RNA interference in crop plants. Curr Opin Biotech 15:139–143

Leegood RC (2007) A welcome diversion from photorespiration. Nat Biotech 25:539–540

Levee V, Garin E, Klimaszewska K, Seguin A (1999) Stable genetic transformation of white pine (*Pinus strobus* L.) after cocultivation of embryogenic tissues with Agrobacterium tumefaciens. Mol Breed 5:429–440

Levin JS, Thompson WF, Csinos AS, Stephenson MG, Weissinger AK (2005) Matrix attachment regions increase the efficiency and stability of RNA-mediated resistance to tomato spotted wilt virus in transgenic tobacco. Trans Res 14:193–206

Linnemann AK, Platts AE, Doggett N, Gluch A, Bode J, Krawetz SA (2007) Genomewide identification of nuclear matrix attachment regions: an analysis of methods. Biochem Soc Trans 35:612–617

Lippman Z, Martienssen R (2004) The role of RNA interference in heterochromatic silencing. Nature 431:364–70

Makarevitch I, Svitashev SK, Somers DA (2003) Complete sequence analysis of transgene loci from plants transformed via microprojectile bombardment. Plant Mol Biol 52:421–432

Marathe R, Smith TH, Anandalakshmi R, Bowman LH, Fagard M, Mourrain P, Vaucheret H, Vance VB (2000) Plant viral suppressors of post-transcriptional silencing do not suppress transcriptional silencing. Plant J 22:51–59

Martienssen RA (2003) Maintenance of heterochromatin by RNA interference of tandem repeats. Nat Genet 35:213–214

Martienssen RA, Zaratiegui M, Goto DB (2005) RNA interference and heterochromatin in the fission yeast *Schizosaccharomyces pombe*. Trends Genet 21:450–456

Maximova S, Miller C, Antunez de Mayolo G, Pishak S, Young A, Guiltinan MJ (2003) Stable transformation of *Theobroma cacao* L. and influence of Matrix Attachment Regions on GFP expression. Plant Cell Rep 21:872–883

May BP, Lippman ZB, Fang Y, Spector DL, Martienssen RA (2005) Differential regulation of strand-specific transcripts from Arabidopsis centromeric satellite repeats. PLoS Genet 1:e79

McGinnis K, Murphy N, Carlson AR, Akula A, Akula C, Basinger H, Carlson M, Hermanson P, Kovacevic N, McGill MA, Seshadri V, Yoyokie J, Cone K, Kaeppler HF, Kaeppler SM, Springer NM (2007) Assessing the efficiency of RNA interference for maize functional genomics. Plant Physiol 143:1441–1451

Meneghini MD, Wu M, Madhani HD (2003) Conserved histone variant H2A.Z protects euchromatin from the ectopic spread of silent heterochromatin. Cell 112:725–736

Meyer P, Kartzke S, Niedenhoff I, Heidmann I, Bussmann K, Saedler H (1988) A genomic DNA segment from *Petunia hybrida* leads to increased transformation frequencies and simple integration patterns. Proc Natl Acad Sci USA 85:8568–8572

Mirkovitch J, Mirault M-E, Laemmli U (1984) Organisation of the higher-order chromatin loop: specific DNA attachment sites on nuclear scaffold. Cell 39:223–232

Mlynarova L, Hricova A, Loonen A, Nap JP (2003) The presence of a chromatin boundary appears to shield a transgene in tobacco from RNA silencing. Plant Cell 15:2203–2217

Mlynarova L, Jansen RC, Conner AJ, Stiekema WJ, Nap JP (1995) The MAR-mediated reduction in position effect can be uncoupled from copy number-dependent expression in transgenic plants. Plant Cell 7:599–609

Mlynarova L, Loonen A, Heldens J, Jansen RC, Keizer P, Stiekema WJ, Nap JP (1994) Reduced position effect in mature transgenic plants conferred by the chicken lysozyme matrix-associated region. Plant Cell 6:417–426

Mlynarova L, Loonen A, Mietkiewska E, Jansen RC, Nap JP (2002) Assembly of two transgenes in an artificial chromatin domain gives highly coordinated expression in tobacco. Genetics 160:727–740

Mourrain P, Beclin C, Elmayan T, Feuerbach F, Godon C, Morel JB, Jouette D, Lacombe AM, Nikic S, Picault N, Remoue K, Sanial M, Vo TA, Vaucheret H (2000) Arabidopsis SGS2 and SGS3 genes are required for posttranscriptional gene silencing and natural virus resistance. Cell 101:533–542

Muller A, Marins M, Kamisugi Y, Meyer P (2002) Analysis of hypermethylation in the RPS element suggests a signal function for short inverted repeats in de novo methylation. Plant Mol Biol 48:383–399

Nagy F, Morelli G, Fraley RT, Rogers SG, Chua NH (1985) Photoregulated expression of a pea *rbcS* gene in leaves of transgenic plants. EMBO J 4:3063–3068

Niessen M, Thiruveedhi K, Rosenkranz R, Kebeish R, Hirsch H-J, Kreuzaler F, Peterhansel C (2007) Mitochondrial glycolate oxidation contributes to photorespiration in higher plants. J Exp Bot 58:2709–2715

Oh SJ, Jeong JS, Kim EH, Yi NR, Yi SI, Jang IC, Kim YS, Suh SC, Nahm BH, Kim JK (2005) Matrix attachment region from the chicken lysozyme locus reduces variability in transgene expression and confers copy number-dependence in transgenic rice plants. Plant Cell Rep 24:145–154

Park YD, Papp I, Moscone EA, Iglesias VA, Vaucheret H, Matzke AJM, Matzke MA (1996) Gene silencing mediated by promoter homology occurs at the level of transcription and results in meiotically heritable alterations in methylation and gene activity. Plant J 9:183–194

Pathak RU, Rangaraj N, Kallappagoudar S, Mishra K, Mishra RK (2007) Boundary element-associated factor 32B connects chromatin domains to the nuclear matrix. Mol Cell Biol 27:4796–4806

Paul AL, Ferl RJ (1998) Higher order chromatin structures in maize and Arabidopsis. Plant Cell 10:1349–1359

Paulson JR, Laemmli UK (1977) The structure of histone-depleted metaphase chromosomes. Cell 12:817–828

Pecinka A, Kato N, Meister A, Probst AV, Schubert I, Lam E (2005) Tandem repetitive transgenes and fluorescent chromatin tags alter local interphase chromosome arrangement in *Arabidopsis thaliana*. J Cell Sci 118:3751–3758

Pederson T (2000) Half a century of "The Nuclear Matrix". Mol Biol Cell 11:799–805

Petersen K, Leah R, Knudsen S, Cameron-Mills V (2002) Matrix attachment regions (MARs) enhance transformation frequencies and reduce variance of transgene expression in barley. Plant Mol Biol 49:45–58

Phi-Van L, von Kries JP, Ostertag W, Stratling WH (1990) The chicken lysozyme 5′ Matrix Attachment Region increases transcription from a heterologous promoter in heterologous cells and dampens position effects on the expression of transfected genes. Mol Cell Biol 10:2302–2307

Poljak L, Seum C, Mattioni T, Laemmli UK (1994) SARs stimulate but do not confer position independent gene expression. Nucleic Acids Res 22:4386–4394

Pwee KH, Webster CI, Gray JC (1994) HMG protein binding to an A/T-rich positive regulatory region of the pea plastocyanin gene promoter. Plant Mol Biol 26:1907–1920

Que QD, Wang HY, English JJ, Jorgensen RA (1997) The frequency and degree of cosuppression by sense chalcone synthase transgenes are dependent on transgene promoter strength and are reduced by premature nonsense codons in the transgene coding sequence. Plant Cell 9:1357–1368

Ragoczy T, Bender MA, Telling A, Byron R, Groudine M (2006) The locus control region is required for association of the murine b-globin locus with engaged transcription factories during erythroid maturation. Genes Dev 20:1447–1457

Richards EJ, Elgin SC (2002) Epigenetic codes for heterochromatin formation and silencing: rounding up the usual suspects. Cell 108:489–500

Robertson D (2004) VIGS vectors for gene silencing: many targets, many tools. Ann Rev Plant Biol 55:495–519

Rozowsky JS, Newburger D, Sayward F, Wu J, Jordan G, Korbel JO, Nagalakshmi U, Yang J, Zheng D, Guigo R, Gingeras TR, Weissman S, Miller P, Snyder M, Gerstein MB (2007) The DART classification of unannotated transcription within the ENCODE regions: associating transcription with known and novel loci. Genome Res 17:732–745

Rudd S, Frisch M, Grote K, Meyers BC, Mayer K, Werner T (2004) Genome-wide in silico mapping of scaffold/matrix attachment regions in Arabidopsis suggests correlation of intragenic scaffold/matrix attachment regions with gene expression. Plant Physiol 135:715–722

Sabo PJ, Kuehn MS, Thurman R, Johnson BE, Johnson EM, Cao H, Yu M, Rosenzweig E, Goldy J, Haydock A, Weaver M, Shafer A, Lee K, Neri F, Humbert R, Singer MA, Richmond TA, Dorschner MO, McArthur M, Hawrylycz M, Green RD, Navas PA, Noble WS, Stamatoyannopoulos JA (2006) Genome-scale mapping of DNase I sensitivity in vivo using tiling DNA microarrays. Nat Methods 3:511–518

Sandhu JS, Webster CI, Gray JC (1998) A/T-rich sequences act as quantitative enhancers of gene expression in transgenic tobacco and potato plants. Plant Mol Biol 37:885–896

Sarcinella E, Zuzarte PC, Lau PNI, Draker R, Cheung P (2007) Monoubiquitylation of H2A.Z distinguishes its association with euchromatin or facultative heterochromatin. Mol Cell Biol 27:6457–6468

Sawasaki T, Takahashi M, Goshima N, Morikawa H (1998) Structures of transgene loci in transgenic Arabidopsis plants obtained by particle bombardment: junction regions can bind to nuclear matrices. Gene 218:27–35

Schoffl F, Schroder G, Kliem M, Rieping M (1993) An SAR-sequence containing 395 bp-DNA fragment mediates enhanced, gene-dosage-correlated expression of a chimaeric heat shock gene in transgenic tobacco plants. Trans Res 2:93–100

Sidorenko L, Bruce W, Maddock S, Tagliani L, Li XG, Daniels M, Peterson T (2003) Functional analysis of two matrix attachment region (MAR) elements in transgenic maize plants. Trans Res 12:137–154

Singh GB, Kramer JA, Krawetz SA (1997) Mathematical model to predict regions of chromatin attachment to the nuclear matrix. Nucleic Acids Res 25:1419–1425

Slatter RE, Dupree P, Gray JC (1991) A scaffold-associated DNA region is located downstream of the pea plastocyanin gene. Plant Cell 3:1239–1250

Soppe WJJ, Jasencakova Z, Houben A, Kakutani T, Meister A, Huang MS, Jacobsen SE, Schubert I, Fransz PF (2002) DNA methylation controls histone H3 lysine 9 methylation and heterochromatin assembly in Arabidopsis. EMBO J 21:6549–6559

Spiker S, Murray MG, Thompson WF (1983) DNase I sensitivity of transcriptionally active genes in intact nuclei and isolated chromatin of plants. Proc Natl Acad Sci USA 80:815–819

Stam M, Mol JNM, Kooter JM (1997) The silence of genes in transgenic plants. Ann Bot 79:3–12

Stief A, Winter DM, Stratling WH, Sippel AE (1989) A nuclear DNA attachment element mediates elevated and position-independent gene activity. Nature 341:343–345

Taddei A, Hediger F, Neumann FR, Gasser SM (2004) The function of nuclear architecture: a genetic approach. Ann Rev Genet 38:305–345

Taddei A, Van Houwe G, Hediger F, Kalck V, Cubizolles F, Schober H, Gasser SM (2006) Nuclear pore association confers optimal expression levels for an inducible yeast gene. Nature 441:774–778

Takano M, Egawa H, Ikeda JE, Wakasa K (1997) The structures of integration sites in transgenic rice. Plant J 11:353–361

Talbert PB, Henikoff S (2006) Spreading of silent chromatin: inaction at a distance. Nat Rev Genet 7:793–803

Tamaru H, Selker EU (2001) A histone H3 methyltransferase controls DNA methylation in *Neurospora crassa*. Nature 414:277–283

Tang G, Galili G (2004) Using RNAi to improve plant nutritional value: from mechanism to application. Trends Biotech 22:463–469

Tariq M, Paszkowski J (2004) DNA and histone methylation in plants. Trends Genet 20:244–251

Tetko IV, Haberer G, Rudd S, Meyers B, Mewes HW, Mayer KF (2006) Spatiotemporal expression control correlates with intragenic scaffold matrix attachment regions (S/MARs) in *Arabidopsis thaliana*. PLoS Comp Biol 2:e21

Thompson WF, Spiker S, Allen GC (2006) Matrix attachment regions. In: Grasser KD (ed) Regulation of transcription in plants. Blackwell, Oxford, pp 350

Ulker B, Allen GC, Thompson WF, Spiker S, Weissinger AK (1999) A tobacco matrix attachment region reduces the loss of transgene expression in the progeny of transgenic tobacco plants. Plant J 18:253–263

Vain P, James VA, Worland B, Snape JW (2002) Transgene behaviour across two generations in a large random population of transgenic rice plants produced by particle bombardment. Theor Appl Genet 105:878–889

Vaistij FE, Jones L, Baulcombe DC (2002) Spreading of RNA targeting and DNA methylation in RNA silencing requires transcription of the target gene and a putative RNA-dependent RNA polymerase. Plant Cell 14:857–867

Valenzuela L, Kamakaka RT (2006) Chromatin Insulators. Ann Rev Genet 40:107–138

Vain P, Afolabi AS, Worland B, Snape JW (2003) Transgene behaviour in populations of rice plants transformed using a new dual binary vector system: pGreen/pSoup. Theor Appl Genet 107:210–217

Van Blokland R, ten Lohuis M, Meyer P (1997) Condensation of chromatin in transcriptional regions of an inactivated plant transgene: evidence for an active role of transcription in gene silencing. Mol Gen Genet 257:1–13

Van der Geest AHM, Hall GE, Spiker S, Hall TC (1994) The β-phaseolin gene is flanked by matrix attachment regions. Plant J 6:413–423

Van Leeuwen W, Mlynarova L, Nap JP, van der Plas LH, van der Krol AR (2001) The effect of MAR elements on variation in spatial and temporal regulation of transgene expression. Plant Mol Biol 47:543–554

Vaucheret H (1993) Identification of a general trans-silencer for 19S and 35S promoters in a transgenic tobacco plant: 90 bp of homology in the promoter sequences are sufficient for trans-inactivation. C R Acad Sci Paris Life Sci 316:1471–1483

Vaucheret H, Elmayan T, Thierry D, van der Geest A, Hall T, Conner AJ, Mlynarova L, Nap JP (1998) Flank matrix attachment regions (MARs) from chicken, bean, yeast or tobacco do not prevent homology-dependent trans-silencing in transgenic tobacco plants. Mol Gen Genet 259:388–392

Vaucheret H, Nussaume L, Palauqui JC, Quillere I, Elmayan T (1997) A transcriptionally active state is required for post-transcriptional silencing (cosuppression) of nitrate reductase host genes and transgenes. Plant Cell 9:1495–1504

Vega-Palas MA, Ferl RJ (1995) The Arabidopsis *Adh* gene exhibits diverse nucleosome arrangements within a small DNase I-sensitive domain. Plant Cell 7:1923–1932

Volpe TA, Kidner C, Hall IM, Teng G, Grewal SI, Martienssen RA (2002) Regulation of heterochromatic silencing and histone H3 lysine-9 methylation by RNAi. Science 297:1833–1837

Washietl S, Pedersen JS, Korbel JO, Stocsits C, Gruber AR, Hackermuller J, Hertel J, Lindemeyer M, Reiche K, Tanzer A, Ucla C, Wyss C, Antonarakis SE, Denoeud F, Lagarde J, Drenkow J, Kapranov P, Gingeras TR, Guigo R, Snyder M, Gerstein MB, Reymond A, Hofacker IL, Stadler PF (2007) Structured RNAs in the ENCODE selected regions of the human genome. Genome Res 17:852–864

Waterhouse PM, Helliwell CA (2003) Exploring plant genomes by RNA-induced gene silencing. Nat Rev Genet 4:29–38

Weintraub H, Groudine M (1976) Chromosomal subunits in active genes have an altered conformation. Science 193:848–856

Wong MM, Cox LK, Chrivia JC (2007) The chromatin remodeling protein, SRCAP, Is critical for deposition of the histone variant H2A.Z at promoters. J Biol Chem 282:26132–26139

Zaratiegui M, Irvine DV, Martienssen RA (2007) Noncoding RNAs and gene silencing. Cell 128:763–776

Zhao T, Palotta M, Langridge P, Prasad M, Graner A, Schulze-Lefert P, Koprek T (2006) Mapped Ds/T-DNA launch pads for functional genomics in barley. Plant J 47:811–826

Zilberman D, Cao X, Johansen LK, Xie Z, Carrington JC, Jacobsen SE (2004) Role of Arabidopsis ARGONAUTE4 in RNA-directed DNA methylation triggered by inverted repeats. Curr Biol 14:1214–1220

Chromatin Domains and Function

Paul Fransz

Abstract The inheritance of biological traits involves not only the transfer of genetic information in the form of DNA, but also epigenetic information. The latter is encrypted in a DNA component, methylation of cytosine residues, and in non-DNA components such as histone modifications, non-histone proteins, and RNA. Chromatin comprises both genetic and epigenetic information. The chromatin state determines which gene will be expressed, at which stage and in which cell. This concept forms the basis for cell differentiation during the development of a eukaryotic organism. It entails a high flexibility of chromatin with respect to molecular composition, biochemical modification, and physical organization. Indeed, chromatin is highly dynamic, showing continuous changes at different levels of organization and occupies different subdomains of the nucleus, each with specific functions. In this chapter, an overview will be given of chromatin states and nuclear domains in which chromatin resides.

1 Introduction

The generation of a large variety of cell types from a single zygotic cell is intrinsic to the inheritance of both genetic and epigenetic information. All cells in a multicellular organism have the same genetic constitution, yet their appearance and function may differ enormously, due to differences in the nuclear program. These differences are established by epigenetic changes during development. In all cells the same DNA sequence directly or indirectly interacts with different combinations of regulatory elements, such as transcription factors, histone modifiers, chromatin remodeling complexes, and other control elements. Consequently, there is a constant change in the composition of molecular visitors at any DNA sequence locus. Frequency, succession, and nature of the visits determine if the chromatin region is

P. Fransz
Swammerdam Institute for Life Sciences, University of Amsterdam,
Amsterdam, The Netherlands
e-mail: p.f.fransz@uva.nl

Plant Cell Monogr, doi:10.1007/7089_2008_23

permissive for the next interaction and eventually change its activity. The dynamic structure of chromatin is therefore fundamental in genome management. The storage of genetic information in the nucleus involves the packing of long chromatin fibers. A fundamental condition is that the information must be retrievable upon request by the regulatory system. Moreover, the genetic information should be available in a cell type and stage-specific manner. This demands a highly efficient packing of chromatin. Some genes, such as household genes, are always transcribed, while other genes are switched off during most of the life time of an organism. In between these extremes, there is a range of gene activity levels. With respect to the control of genetic information, three hierarchical levels can be considered (van Driel et al. 2003). These comprise (1) the DNA level, i.e. the linear organization of transcription units and *cis*-regulatory sequences, (2) the chromatin level, which involves the eukaryotic interface between DNA sequence and the functional state of the locus and (3) the nuclear level, which includes the dynamic three-dimensional organization of the genome. The latter facilitates a proper environment for chromatin changes, but is also the result of variable chromatin states. Hence, the nuclear architecture is dependent on the functional state of chromatin, which in turn is controlled by histone modifications in cross-talk with DNA methylation and RNAi (reviewed by Fransz et al. 2006). Sequence specificity, however, is established by *cis*-regulatory elements (enhancers, boundary elements, promotors), repeat organization, transcription factors, and siRNA. This chapter focuses on the relation between the epigenetic code of chromatin and the organization of subnuclear domains.

2   Chromatin

2.1   *Nucleosome Folding*

Chromatin consists of a constantly changing complex of DNA associated with proteins and RNA. The basic unit of chromatin is formed by the nucleosome, an octamere configuration of two times four core histones, H2A, H2B, H3, and H4, around which 146 pairs of the DNA helix are wrapped in ~1.7 superhelical turns. Adjacent nucleosomes are connected via linker DNA stretches. An array of nucleosomes gives the typical "beads on a string" appearance, which has been visualized by electron microscopy (Oudet et al. 1975) and atomic force microscopy (Jimenez-Garcia and Fragoso-Soriano 2000; Leuba et al. 1994). However, we do not know if this level of organization exists in living cells, since these observations were made from an artificial situation. In addition, in living cells the association of linker histones (histone H1) will generally lead to a more compacted state of chromatin. Moreover, non-histone protein complexes will decorate the "naked" nucleosome fiber and generate higher levels of chromatin organization. The first compaction level of chromatin in living cells is likely the 30-nm fiber (Finch and Klug 1976; Thoma et al. 1979). The thickness of the 30-nm fibers may show variation depending on the average length of the linker DNA. Two helical models have been proposed for the 30-nm fiber. The one-start helix (solenoid) is linearly arranged, whereas the two-start helix (zig-zag) is arranged as two nucleosomal rows in such a way that following nucleosomes alternate from one to the other row. Linker histones are located in the interior of the fiber. The zig-zag model implies a double helical structure in which alternative nucleosomes become interacting partners. Recent experimental data, including crystal structure analysis, reveal two stacks of nucleosomes, which strongly favors the two-start model (Dorigo et al. 2004; Schalch et al. 2005). The variable topology of the two-start model not only allows maximum packing density, but also facilitates transitions between folded and unfolded chromatin (Wu et al. 2007). A critical factor in the two-start helix is the length of the linker DNA, which is generally short in active chromatin and long in tightly packed chromatin such as in spermatids. Interaction between core histones from adjacent nucleosomes in different rows has been shown for histone H2A and the N-terminal tail of histone H4 (Luger et al. 1997; Schalch et al. 2005). Moreover, acetylation of the histone H4 tail affects the formation of the 30-nm fiber (Shogren-Knaak et al. 2006). Consequently, histone modifications can directly affect the compaction of the chromatin fiber.

2.2   Histone Code at the Surface of the Nucleosome

Histone folding enables histone–histone interactions and the binding to the DNA helix, while histone amino-terminal tails pass over and between the openings of the DNA superhelix to contact nearby molecules (Luger et al. 1997). The typical folding properties of histones are therefore essential for histone modification and consequently for higher-order structure of chromatin fiber and the accessibility of DNA. Since the early 1970s it has been known that core histones can be covalently modified at their tails via phosphorylation, methylation, acetylation, and ubiquitylation. But it was only during the last decade that their crucial role in gene regulation became evident. At the turn of the century the histone code was proposed to explicate the crucial role of histone modifications in epigenetic gene regulation (Strahl and Allis 2000; Turner 2000). It comprises that specific modifications at the histone tails establish a molecular code at the nucleosome surface that is recognized by effector proteins to induce downstream events. By that time several modifications were known, of which the best studied was acetylation of lysine residues. A major breakthrough came later in 2000 when the group of Jenuwein demonstrated that the SET (Suvar3-9, Enhancer-of-zeste, Trithorax) domain of SUVH39, the human homolog of Su(var)3-9, the dominant suppressor of position effect variegation in *Drosophila*, is responsible for histone H3 methyltransferase (HMT) activity (Rea et al. 2000). This SET domain is highly conserved among yeast, animals, and plants. The authors also demonstrated interdependence of site-specific histone tail modifications. Acetylation of H3K9 and phosphorylation at H3S10 both prevented H3K9 methylation, whereas phosphorylation of H3S10 was inhibited when H3 was already methylated at lysine 9. A positive interaction was found between acetylated H3K14 and H3S10 phosphorylation. The histone code hypothesis was tested by Agalioti et al. (2002), who mutated the lysine residues in histones H3 and H4 that were known to be acetylated during a viral infection. They discovered a biochemical cascade for gene activation at the IFN-b gene through ordered recruitment of transcription complexes by acetylation of specific lysines. H4K8Ac recruits SWI/SNF, while H3K9Ac and H3K14Ac recruit TFIID. The authors provided evidence that information contained in the DNA sequence (i.e. enhancer) is transferred to the histone tails by the formation of molecular sites that can interact with transcription complexes.

In the past few years many more modified residues have been established including acetylation, methylation, phosphorylation, ubiquitylation, and sumoylation (for reviews see Berger 2007; Margueron et al. 2005), all of which contribute to the variation of the molecular code laid down at the surface of the nucleosome. The number of modulations of the histone code is extended by the level of post-transcriptional modifications. For example, lysines can become mono-, di-, or tri-methylated, while acetylation can occur up to the tetra-level. It follows that a group of residues in one tail can increase the variation in the histone code. Indeed, using chromatography and high-resolution tandem mass spectrometry Garcia et al. (2007) characterized over 150 different modified isoforms of human histone H3.2, implicating as many distinct "histone codes" for histone H3.2. The authors further predicted the existence of many hundreds of differently modified forms of histone H3 variant in human cells.

2.3   Histone Variants Expand the Histone Code

In addition to histone modification, eukaryotes have developed more tools to modify the epigenetic state of chromatin. Canonical histones can be replaced by variant histones, which have a different amino-acid sequence. Accordingly, they have different biophysical characteristics, which alter the properties of nucleosomes. Unlike the major histones, histone variants are generally deposited into nucleosomes during the G1 or G2 phase, rather than during DNA replication. Furthermore, histone variants localize to specific regions of the chromosome, giving rise to distinct changes in chromatin function. For example, the canonical histone H3 (H3.1) is replaced at the centromere region by the centromere-specific variant CENH3 (CENP-A), which directs the assembly of the kinetochore ensuring proper segregation of the sister chromatids to the daughter cells (Ahmad and Henikoff 2002; Blower et al. 2002). In fact, the functional centromere is determined by the presence of CENH3 rather than the DNA sequence. Moreover, in a recent study it is suggested that differential cytosine methylation of the functional centromere and flanking regions plays an important role in the binding of CENH3 to the centromere core (Zhang et al. 2008). Another variant of histone H3 is histone H3.3. Since this variant is enriched in euchromatin at active genes, it is supposed to be incorporated into active chromatin during transcription (McKittrick et al. 2004). Histones H2A.Z and H2A.X are variants of histone H2A, present in most eukaryotes, while MacroH2A and H2A-Bbd have only been found in animals. Histone H2A. X plays a role in double-strand break repair (Li et al. 2005), whereas H2A.Z is functional in different processes including gene activation and repression (Bruce et al. 2005; Meneghini et al. 2003; Rangasamy et al. 2003). Using reversed-phase chromatography and tandem mass spectrometry Bergmuller et al. (2007) present the first detailed characterization of H2B-variants isolated from *Arabidopsis*. In addition, they discovered several novel modifications, including methylation of N-terminal alanine residues. The histone variants further increase the number of combinations in the histone code.

2.4 DNA Methylation Provide More Epigenetic Variation

Apart from the histone code there is a DNA methylation code, which is based on the methylation of cytosine residues at position 5, located in the major groove of the DNA helix. In contrast to histone modifications and histone variants, which leads to either activation or repression, DNA methylation is mainly associated with repression of genes and transposon repeats. Plants can methylate cytosines in the symmetric CpG or CpNpG context, but also at asymmetrical CpNpNp sites (reviewed in Bender 2004; Fransz et al. 2006). In comparison, mammals only have CpG methylation, except in embryonic stem cells where non-CpG methylation is prevalent (Ramsahoye et al. 2000). *Drosophila* and yeast on the other hand hardly have any DNA methylation or no DNA methylation at all. DNA methylation is therefore not as universal as histone modification. DNA methylation in *Arabidopsis* is established by four different DNA methyltransferases with partially separated and overlapping functions. DRM1 and DRM2 (domains rearranged methyltransferase) are de novo methyltransferases, responsible for virtually all de novo DNA methylation. MET1 (methyltransferase) is a maintenance DNA methyltransferase, responsible for CpG methylation at hemimethylated sites. CMT3 (chromomethyltransferase) is involved in de novo methylation at non-CpG positions and also maintains CpNpG methylation. DNA methylation may affect gene activity in a direct or indirect way. The presence of methylated cytosines likely alters the hydrophobic environment in the major groove, thereby affecting the binding of transcription factors to a regulatory sequence and inhibiting transcription. Indirectly, methylated cytosines can be recognized by MBD (methyl-CpG-binding domain) proteins, which can recruit co-repressors such as histone deacetylases or histone methyltransferase leading to gene silencing (see below). Furthermore, the repressive control of DNA methylation over methylated H3K4 has been demonstrated at intragenic sites (Lorincz et al. 2004; Okitsu and Hsieh 2007). In addition, it has been demonstrated that the histone code can direct DNA methylation in yeast (Tamaru and Selker 2001), mammals (Lehnertz et al. 2003), and plants (Jackson et al. 2002), indicating that the two codes can cross-talk and reinforce each other.

The selective manner of cytosine methylation, the replacement of canonical histones by histone variants and the histone code together form the epigenetic code.

2.5   Reading the Epigenetic Code

Histone modifications, histone variants, and methylated cytosines mark the genomic sites that require attention from the regulatory system to direct effector molecules. In order to affect DNA processes (transcription, replication, and repair), the effectors must change or maintain the functional state of chromatin via catalytic activity and/or a physical change, such as folding. Consequently, to give the epigenetic code functional significance the epigenetic marks need to be recognized. At least three main classes of chromatin binding domains (CBD) are instrumental in reading the epigenetic code. The methyl-CpG-binding domain (MBD) interacts with methylated cytosines, the chromodomain recognizes methylated histones and the bromodomain binds to acetylated histones. Effector proteins are either recruited to their target by CBD proteins or they may contain a CBD. The combination of effector domains and CBDs provides an enormous variation in the interpretation of the epigenetic code (Seet et al. 2006). In addition, among members of a class there is variation in target specificity and target affinity. For example, all bromodomains can interact with active chromatin due to their affinity to acetylated histones. However, preferential binding to specific targets has been reported. In a pull-down assay with bromodomains of seven different co-activators, it was found that the bromodomain of SWI2/SNF2, the ATP-dependent nucleosome remodeller, interacts with several acetylated histones H3 and H4. In contrast, the bromodomain of the histone deacetylase in the Gcn5 complex interacts only with acetylated H3 and with tetra-acetylated H4 tails, while the bromodomain of the histone acetyltransferase Spt7 in the SAGA complex interacts weakly with acetylated H3 and not with acetylated H4 at all (Hassan et al. 2007).

A similar difference in the specificity of targeting was found for chromodomains. The chromodomain of the Heterochromatin Protein 1 (HP1) binds specifically to methylated H3K9 creating a heterochromatic state for repeat regions, while the chromodomain of the Polycomb (Pc) protein, involved in maintaining repression of homeotic genes, interacts with methylated H3K27. By swapping the chromodomain regions of HP1 and Pc Fischle et al. (2003) demonstrate that the chromodomain is sufficient to switch the nuclear pattern of these chromatin proteins. The experiment illustrates the essential role of these domains in reading the histone code. Moreover, even within the family of Pc proteins, the chromodomain, despite a high degree of conservation, displays a remarkable difference in affinity for H3K27me3 and H3K9me3 (Bernstein et al. 2006). In this context it is remarkable that the plant homolog of HP1, Like Heterochromatin Protein (LHP1), has affinity for both methylated H3K9 and H3K27 (Zhang et al. 2007). It has been proposed by several authors that LHP1 does not have the same role in plants as HP1 has in animals (Jackson et al. 2002; Johnson et al. 2004; Libault et al. 2005;

Lindroth et al. 2004; Nakahigashi et al. 2005). Strikingly, LHP1 co-localizes with H3K27 genome-wide (Turck et al. 2007; Zhang et al. 2007), which strongly supports the view that LHP1 functions as a Pc protein, because of its chromodomain (see below).

The conserved region of proteins binding to methylated cytosine is the methyl-CpG-binding domain, which shows high similarity between plant and human sequences. The *Arabidopsis* genome contains 12 MDB proteins divided into seven subclasses (Berg et al. 2003; Scebba et al. 2003; Zemach and Grafi 2003). Not all of these proteins possessing an MBD protein bind to methylated cytosine. Similar to the bromodomain and chromodomain, MBDs show target specificity. AtMBD6 and AtMBD7 bind only to methylated CpG targets, while AtMBD5 can bind to both methylated cytosines in CpG and CpNpN contexts. The classes containing these three MBD proteins are unique to dicots (Springer and Kaeppler 2005). Considering that DNA methyltransferases show significant conservation between plants and animals, while MBD proteins are only conserved within the MBD motif, eukaryotes may have evolved different mechanisms to interpret the DNA methylation code.

Recently, the SRA (SET and RING associated) domain, a non-MBD domain, has been identified to bind to methylated DNA (Johnson et al. 2007; Woo et al. 2007). This motif was discovered in several histone methyltransferases (HMTs, Baumbusch et al. 2001). The SRA domain in SUVH4 (KRYPTONITE) is required to recruit the methyltransferase to methylated DNA in different sequence contexts (Johnson et al. 2007). The SRA domain in the Variant In Methylation 1 (VIM1) protein, belonging to a family that is implicated in chromatin modification, is required for maintaining centromeric heterochromatin (Liu et al. 2007; Woo et al. 2007). Taken together, the interpretation of the epigenetic code is carried out by CBD-containing proteins that further expand the epigenetic language by combining multiple functional domains for recognition, binding, folding, or catalyzing targets.

3   Nuclear Organization of Chromatin

3.1   Different Levels of Chromatin Compaction in the Interphase Nucleus

The epigenetic code dictates the chromatin state and alters the nuclear architecture. Vice versa, the organization of the nucleus into subdomains facilitates a proper environment for chromatin changes. Since there are no membrane-bound domains inside the nucleus, molecular components can freely travel from one chromatin domain to another. This is demonstrated for large molecules up to several hundreds of kilodaltons (Verschure et al. 2003). Yet the nucleus displays a differential organization in which few morphological features determine its microscopic appearance: nucleolus, heterochromatin domains, and chromosome territories. Apparently there is a preference of molecular components for specific subdomains.

Nuclear subdomains have been described from the beginning of the previous century, using typical dyes such as hematoxylin, Feulgen, and aceto carmine for nucleic acids and eosin for proteins. These studies enabled Heitz in 1928 to distinguish two classes of chromatin: euchromatin and the more condensed heterochromatin. The latter was further subdivided into a permanent or constitutive type of heterochromatin and a transient or facultative type of heterochromatin (Brown 1966). An overview of typical features of heterochromatin is shown in Table 1. However, since there is no absolute parameter for heterochromatin all characteristics should be considered relative. For example, the condensed nature of heterochromatin becomes only obvious if it is flanked by less condensed euchromatin. But even within the two chromatin types there can be large variation in compaction. If we assume the genome of *Arabidopsis* to be approximately 160 Mb (Bennett et al. 2003) partitioned over five chromosomes with an average size of 2 μm during metaphase (Maluszynska and Heslop-Harrison 1991), then the linear compaction at this phase is about 16 Mb per μm. During replication, parts of the DNA helix are unwound and the compaction degree may be close to that of the B-configuration of DNA. According to the Watson and Crick model this is about 2.93 kb per μm. In between the compact metaphase chromosome and the open DNA helix during replication there are several levels of chromatin compaction. For instance, in pachytene cells the heterochromatin nucleolar organizing region (NOR) measures 2.3 Mb per μm, while the heterochromatic knob hk4S has a condensation degree of

Table 1 General characteristics of heterochromatin

Cytological:

• Dark regions in phase contrast microscopy
• Intensely stained regions with general chromatin dyes
• C-band positive
• G-band positive (not in plants)

Molecular:

• Chromosome segments containing abundant (tandemly) repeated DNA and enriched in transposons

Biochemical:

• A DNA–protein complex that is relatively insensitive to DNase I
• Enriched in methylated DNA (not in *Drosophila* and yeast)
• Enriched in histones H3 methylated at position lysine 9
• Enriched in heterochromatin protein HP1 (not in *Arabidopsis*)

Physical:

• Chromatin with regularly arranged nucleosomes
• Condensed chromatin

Functional:

• Largely inactive in transcription
• Late-replicating
• Rarely involved in meiotic recombination

Table 2 Different compaction degrees in euchromatin and heterochromatin

Species	Cell	Chromosome region	Length (mm)	Size (kb)	Compaction (kb/mm)
Arabidopsis	Pachytene	Total complement	331.2	160,000	483
		NOR	3.3	7,500	2,301
		180-bp repeat	8.3	5,000	600
		Heterochromatic knob 4S	1.0	700	700
		Transposons + 5S	12.9	18,500	1,434
		Euchromatin	307.4	120,000	390
	Metaphase	Total complement	10.0	160,000	16,000
	Interphase	Heterochromatic knob 4S	0.7	700	1,000
		Short arm 4S	6.0	2,800	467
		YAC (7C3)	2.6	470	181
		BAC contig (T19B17-T27D20)	3.4	185	55
Tomato	Pachytene	Total complement	483.0	950,000	1,967
	Metaphase	Total complement	35.0	950,000	27,143
Medicago	Pachytene	Total complement	406.0	500,000	1,232
	Metaphase	Total complement	20.3	500,000	24,631
	Pachytene	BAC contig (58F01-59K07)	0.5	150	300

Data are based on FISH studies in *Arabidopsis* (Fransz et al. 1998, 2000, 2002), tomato (Zhong et al. 1998) and *Medicago* (Kulikova et al. 2001)

700 kb per μm (Fransz et al. 2000). Moreover, across different species the level of chromatin compaction differs dramatically. Table 2 presents a number of measured condensation degrees for several regions of the *Arabidopsis* genome in different cell types. In comparison, compaction of the 30-nm fiber of chromatin varies between 70 kb and 160 kb per μm depending on linker length (Robinson et al. 2006). From these data we infer that in *Arabidopsis* the lower levels of euchromatin compaction during interphase correspond with the level of the 30-nm fiber.

3.2 Chromosome Territory

Heterochromatin and euchromatin segments are visible along the linear structure of the chromosomes and in the three-dimensional organization of the chromosomes during interphase. Obviously, the linear and three-dimensional configurations of chromatin segments are related to each other. This is illustrated by the nuclei of three species with different genome size and distribution of heterochromatin segments (Fig. 1a). Plants with large chromosomes such as barley (1C = 4,900 Mb) and onion (1C = 16,000 Mb) have highly condensed chromatin all over the chromosome arms (de Jong et al. 1999), while in tomato (1C = 950 Mbp) the arms contain distinct euchromatic segments and large heterochromatic regions at the distal end, interstitially and especially in the proximal part (Zhong et al. 1998). In contrast, the chromosome arms in *Arabidopsis* (1C = 160 Mbp) are largely euchromatin, with relatively small pericentromeric heterochromatin (Fransz et al. 1998). The hetero-chromatin–euchromatin proportion is reflected in the nuclear morphology. In a barley nucleus there is no distinguishable euchromatin area, while in tomato and *Arabidopsis* both euchromatin and heterochromatin domains are clearly visible (Fig. 1b–d). Also the organization of individual chromosomes may reflect the linear arrangement of chromatin segments along the chromosome. In tomato the distal

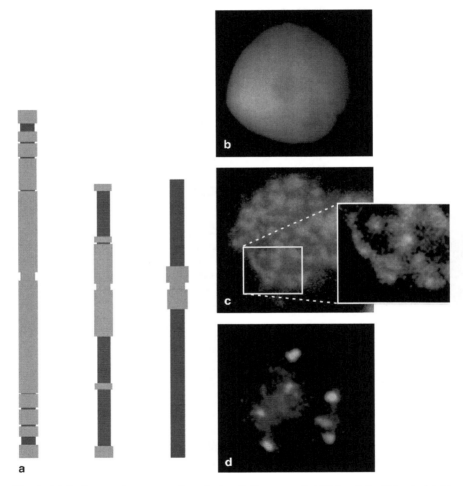

Fig. 1 (**a**) Pachytene chromosome from barley (*left*), tomato (*middle*) and *Arabidopsis* (*right*), showing the distribution of heterochromatin segments (*light grey*) along the chromosome arms (*dark grey*). (**b–d**) Microscopical images of DAPI-stained interphase nuclei from barley (**b**), tomato (**c**), and *Arabidopsis* (**d**), showing the distribution of heterochromatin domains (*bright domains*). The magnification of the tomato nucleus shows telomere signals (*bright spots*) at the borders of the heterochromatic islands

heterochromatin segments form islands of heterochromatin with other proximal heterochromatin. This is shown by the presence of telomere sequences at the border of the heterochromatin domains (Fig. 1c). In contrast, the large chromosomes remain in a Rabl orientation with telomeres and centromeres at opposite nuclear poles (Jasencakova et al. 2001). Even when chromatin is activated after inducing a hypomethylated situation, wheat chromosomes remain in a Rabl orientation (Santos et al. 2002). *Arabidopsis* does not have distal heterochromatin segments, apart from chromosome arms 2S and 4S, which bear the NOR. Here, the pericentromeric heterochromatin occupies peripheral positions, while the distal chromosome ends flank the nucleolus (Fransz et al. 2002). Each chromosome consists of a heterochromatic domain or chromocenter from which euchromatic loops emanate. Together they form a chromosome territory. On the basis of chromosome painting studies it is generally accepted that interphase chromosomes occupy discrete territories (Cremer and Cremer 2001). In *Arabidopsis*, differential painting revealed a random association of chromosome territories, except for the NOR-bearing chromosomes (#2 and #4), which are generally attached to the nucleolus. However, a significant percentage of allelic sequences appear in close proximity, suggesting not only intermingling of chromosome territories, but also an opportunity for homologous recombination (Pecinka et al. 2004). Indeed, cytogenetic analysis of interphase nuclei revealed frequent association of heterochromatin domains, in particular of the NORs from chromosome 2 (Fransz et al. 2002). In *Arabidopsis*, the average chromosome territory comprises 25 Mb of DNA with 5,200 genes (*Arabidopsis* Genome Initiative 2000). In comparison, the average human chromosome territory is five times larger (130 Mb) but contains only 1,700 (Cremer and Cremer 2001; Human Genome Sequencing 2004). These figures illustrate the differences in gene density and chromatin compaction between different species.

3.3 Chromocenters

3.3.1 Chromocenter Appearance

When heterochromatin clusters into conspicuous, discrete nuclear domains they are called chromocenters. The most striking feature is the sharp transition in appearance from euchromatin to the condensed, heterochromatic chromocenter. Chromocenters in plants were described in 1907 by Laibach, who called them "Chromatin Körnchen" (Laibach 1907). Laibach observed chromocenter numbers per nucleus, from which he could deduce the number of chromosomes. It enabled him to establish the haploid chromosome number of *Arabidopsis thaliana* (n = 5), which by that time was the smallest odd number of chromosomes. Although all eukaryotic organisms contain condensed heterochromatin, not all species display chromocenters. For example, human cells do not have chromocenters. Why some species show distinct chromocenters and others do not is unclear. No chromocenters have been observed in plant species with a large genome (>3,000 Mb) such as barley, wheat, onion, *N. tabacum* or field bean, suggesting that genome size is a limiting factor in the formation of chromocenters. This is also suggested by a study of chromocenters in 67 plant species, all having a relatively small genome (Ceccarelli et al. 1998). In addition, Houben et al. (2003) investigated the nuclear distribution of methylated histone H3K9 (H3K9me2) in 24 plant species and found two different patterns dependent on the genome size. Species with a small genome (1C < 500 Mb) displayed a chromocenter-like pattern, in which H3K9me2 was restricted to constitutive heterochromatin, while species with larger genomes showed a uniform distribution of the epigenetic mark. Yet, genome size is probably not the only factor to determine the appearance of chromocenters. For example, although the genomes of mouse and human are equally large and share many genetic features, the morphology of their nuclei differs. Where mouse nuclei show 20–80 chromocenters (Guenatri et al. 2004), human nuclei show a gradual transition between euchromatin and heterochromatin. Similarly, tomato with a genome size of 950 Mb, which is about two times larger than that of rice (490 Mb), contains large islands of condensed chromatin, whereas rice nuclei show faint or diffuse heterochromatin regions (Houben et al. 2003; Ohmido et al. 2001). In fact, there are no reports of chromocenters in cereals. It is likely that chromocenter appearance, and thus chromosome organization in interphase, is not only related to the genome size but also to the linear organization of chromosomes. The nature of this relation, however, is unknown. In this context it should be noted that humans appear to have stopped accumulating repeated DNA over 50 million years ago, whereas there seems to be no such decline in mouse (http://www.ornl.gov/sci/techresources/Human_Genome). This might explain some of the differences between human and mouse, including the formation of chromocenters.

3.3.2   Chromocenters Contain Major Repeats

Chromocenters generally correspond to constitutive heterochromatin, which stains positively for C-banding. The main genetic components of chromocenters are tandem repeats and dispersed transposable elements. For example, the mouse genome contains two classes of repeat sequences that associate with chromocenters. Major satellite repeats of mouse are A/T-rich and comprise more than 10,000 copies of a 234-bp unit per pericentric region, while the centric minor satellite repeats consists of tandem arrays of ca. 2,000 copies of 123 bp (Martens et al. 2005; Waterston et al. 2002). The minor satellite repeats and the much larger major satellite repeats, together form constitutive heterochromatin. It has been proposed that clustering of compact chromatin into chromocenters is generated by ectopic pairing of repetitive sequences (Comings 1980; Manuelidis 1990). This idea was supported by the fact that chromocenters are formed by the aggregation of major satellite repeats with Heterochromatin Protein 1, whereas the minor satellite repeats formed small separate entities associated with centromere proteins at the periphery of the chromocenter (Guenatri et al. 2004). Towards mitosis the number of minor satellites per chromocenter decreases and the number of chromocenters increases. The cohesion of major satellite repeats in chromocenters reflects the attachment of sister chromatids until separation during anaphase, pointing at a function of heterochromatin in the proper segregation during mitosis.

In *Arabidopsis*, all major tandem repeats are located in chromocenters (Fransz et al. 2002). These include the 45S rDNA gene repeats, the 180-bp centromeric repeat, and the 5S rDNA gene repeats. Whereas all chromocenters contain the 180-bp repeat, the 45S rDNA genes are only present in NOR-chromocenters of chromosomes #2 and #4. The 45S rDNA, which span 3–4 Mb per locus (Copenhaver and Pikaard 1996), form the longest tandem repeat arrays. The length of the smaller 5S rDNA arrays is 100–300 kb per locus (Cloix et al. 2000; The Arabidopsis Genome Initiative 2000) and is mapped to the pericentromeres of chromosomes #3, #4, and #5 in the accession Columbia (Fransz et al. 1998). In addition to the tandem repeats the *Arabidopsis* chromocenters are highly enriched in dispersed transposon repeats, which map to all pericentric regions. Hence, apart from the 5S and 45S rDNA loci the chromosomes do not have major chromosome-specific repeats. Close examination of chromosome territories, painted with gene-rich BAC DNA probes revealed that virtually all genes, active and inactive, localize outside the chromocenters in euchromatin (Fransz et al. 2006).

3.3.3 Chromocenters Contain Epigenetic Marks for Repression

Repeat sequences, including the majority of the ribosomal repeats are targets for epigenetic silencing. Consequently, chromocenters in plants and animals are labeled with DNA methylation, MBD proteins and histone H3K9 methylation, whereas histone acetylation is practically absent. However, there are some differences in the epigenetic markers in chromocenters between plants and animals. Chromocenters in mouse and *Drosophila* contain trimethylated H3K9 and HP1. In plants, however, chromocenters are decorated with dimethylated H3K9 (Houben et al. 2003; Jasencakova et al. 2003), but not with trimethylated H3K9 (Fig. 2), while LHP1, the homolog of HP1, was detected in euchromatin but not in chromocenters

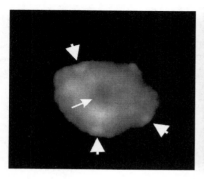 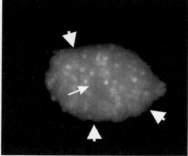

Fig. 2 Immunolabeling of trimethylated H3K9 (*right*) in *Arabidopsis* nucleus counterstained with DAPI (*left*)

(Libault et al. 2005; Nakahigashi et al. 2005). Also for histone H4K20 methylation the chromocenters in *Arabidopsis* and mouse show opposite patterns (Naumann et al. 2005; Schotta et al. 2004). Where monomethylation is high in *Arabidopsis*, this mark is low in mouse. Vice versa, trimethylation of H4K20 is low in chromocenters of *Arabidopsis*, but high in the heterochromatin domains of the mouse (Table 3). It is not clear why chromocenters of *Arabidopsis* and animals have different epigenetic labels. It is possible that plants and animals have developed different mechanisms to interpret the epigenetic code. For example, methylation of H3K27 is a mark for facultative heterochromatin in mammals to silence genes in association with the Polycomb proteins that maintain the repressed state. In mouse and *Arabidopsis* trimethylated H3K27 is localized outside chromocenters. Heterochromatin Protein 1 which recognizes methylated H3K9 accumulates in mouse chromocenters. In plants, however, the homologous LHP1 binds to methylated H3K27 and is found in euchromatic regions. This supports the idea that in

Table 3 Localization of epigenetic marks and effector proteins in nuclear domains

Epigenetic mark or effector protein		Euchromatin	Chromocenter NOR	Chromocenter other	Nucleolus	Refs.[a]
H3K9	Mono	Low	High	High	Low	1
	di	Low	High	High	Low	1,2,3
	Tri	High	Low	Low	Low	1,4
H3K27	Mono	Low	High	High	Low	1,5
	di	Low	High	High	Low	1,5
	tri	High	Low	Low	Low	1,5
H4K20	mono	Low	High	High	Low	1
	di	High	Low	Low	Low	1
	tri	High	Low	Low	Low	1
H3K4	di	High	Low	Low	Low	1,2,3
H4Ac5		High	Low	Low	Low	2
H4Ac8		High	Low	Low	Low	2
H4Ac12		High	Low	Low	Low	2,4
H4Ac16		High	High in root tip	High/low	Low	2,3
H3Ac9		High	High in root tip	Low	Low	2,3
H3K9/18		High	Low	Low	Low	2,3
H1		High	High	High	Low	4
5m-C		Low	High	High	Low	1,2,3,6
AtMBD2	no C	High	Low	Low	Low	7
AtMBD5	CpG, CpNpN	Low	High	Low	Low	7
AtMBD6	CpG	Low	High	Low	Low	7
AtMBD7	CpG	Low	High	High	Low	7
VIM1		High	High	High	Low	8
DDM1		Medium	High	High	Medium	7
HDA6		Medium	Low	Low	High	9

[a] Refs.: 1, Nauman et al. 2005; 2, Soppe et al. 2002; 3, Jasencakova et al. 2003; 4, this paper; 5, Mathieu et al. 2005; 6, Fransz et al. 2002; 7, Zemach et al. 2005; 8, Woo et al. 2007; 9, Earley et al. 2006

plants LHP1 has a function in gene repression that involves Polycomb Respressive Complexes (PRCs). Indeed, LHP1 is required to silence the gene of Flowering Locus C (FLC), a MADS box protein, involved in flowering time in a cooperative manner with PRC proteins (Finnegan and Dennis 2007; Mylne et al. 2006). Further support came from an earlier study in which LHP1 was shown to repress several floral homeotic genes (Kotake et al. 2003). The target specificity of LHP1 may be embedded in the chromodomain. In this context it is interesting to note that introduction of the *Drosophila* HP1 into *Arabidopsis* leads to accumulation in the chromocenters (Naumann et al. 2005). Furthermore, it is known from mouse studies that HP1 recruits SUV4-20 HMTase to heterochromatin (Schotta et al. 2004). This may explain why H4K20 trimethylation in *Arabidopsis* is in gene-rich euchromatin, in contrast to mouse. If LHP1 binds to H3K27 in gene-rich euchromatin it may recruit the HMTase that methylates H4K20.

3.3.4 Chromocenter Compaction

The shape and size of chromocenters have long been thought to vary during embryogenesis, but only marginally at later developmental stages. For example, heterochromatin in *Drosophila* is reduced at early stages of development, while the aggregation of heterochromatin domains in *Drosophila* was monitored especially in salivary gland cells. Constitutive heterochromatin is defined as compact chromatin during all phases of the cell cycle. However, in some gene silencing mutants of *Arabidopsis* the level of chromatin compaction is reduced due to dislocation of pericentric repeats away from the chromocenter (Fransz et al. 2006; Soppe et al. 2002). Recently two studies in *Arabidopsis* have revealed a dramatic reduction of compact chromatin caused by developmental and environmental cues. In protoplasts all chromocenters except for a few NOR chromocenters have disappeared due to unfolding of the repeat regions (Tessadori et al. 2007a). Even the centromeric repeats showed a spectacular decondensation. Strikingly, the process appeared reversible. Upon culturing, chromocenters reassembled, following a sequential process in which long tandem repeat arrays were the first sequences to condense. Similarly, during the floral transition leaf nuclei undergo a reduction in chromocenter compaction, which is reversed upon the appearance of the first flower buds (Tessadori et al. 2007b). Apparently, constitutive heterochromatin is not as permanently compact as we assumed.

3.4 The Nucleolus

3.4.1 Nucleolar Organization

The nucleolus is known as the large subnuclear compartment in which ribosomal genes are transcribed and the transcripts processed to ribosomal subunits. Although first described by Gabriel Valentin about 200 years ago, we are acquainted with the function of the nucleolus in ribosome synthesis only from the second half of the previous century. Ultrastructural analysis of the nucleolus revealed several sub-domains including fibrillar centers (FCs), dense fibrillar components (DFCs), and granular components (GCs), names that refer to their appearance under the electron microscope. Transcription of ribosomal genes occurs at the border between the FC and DFC, while pre-rRNA molecules are processed in the DFC. In the GC region the assembly of ribosome subunits is completed, after which the subunits are trans-ported to the cytoplasm (Tschochner and Hurt 2003). The number of ribosomal genes per haploid genome varies from several hundreds up to many thousands depending on the organism (Rogers and Bendich 1987). The genes are clustered in a special locus, known as the nucleolar organizing region (NOR), which was first discovered by McClintock in 1934.

Light microscopical studies with in situ hybridization using rDNA probes revealed hybridization signals scattered all over the nucleolar region, indicating highly decondensed chromatin in the nucleolus (Montijn et al. 1998) and elevated transcription activity. Indeed, ribosomal rDNA is highly expressed, constituting up to 80% of the total RNA in a cell (Jacob and Ghosh 1999). However, the number of actively transcribed rDNA genes is remarkably small. By far the majority of the ribosomal genes (50–90%) remain inactive (Grummt 2003) and form condensed domains flanking the nucleolus. Active rDNA genes are associated with RNA pol I and respond positive to silver staining. Its chromatin is ten times less condensed than in flanking chromosomal regions (Boisvert et al. 2007). Moreover, during mitosis they are still decondensed forming the so-called secondary constriction.

3.4.2 Regulation of Ribosomal Gene Transcription

The regulation of rDNA gene expression comprises the control of hundreds of ribosomal gene copies. The transcription activity of ribosomal genes can vary greatly among cell types and between different developmental stages. For example, transcription of rDNA in growing cells is higher than in non-growing cells (Grummt 2003). However, a higher copy number of rDNA genes does not result in higher rDNA expression (Flavell 1986; Muscarella et al. 1985). Apparently, there is a mechanism that knows how many rDNA genes need to be "on" at a certain time and in a certain cell. The underlying molecular mechanism controlling the on/off state is not known. It is well documented that epigenetic mechanisms are involved in the control of rDNA transcription. These data show a significant correlation between rDNA activity, DNase I accessibility, DNA methylation and histone modification (reviewed by Preuss and Pikaard 2007). In addition, immunodetection of methyl-ated DNA in differentiated and in proliferating cells showed strong signals at the condensed NOR domains and pericentric heterochromatin, whereas the nucleolus remained empty (Fransz et al. 2002; Naumann et al. 2005), suggesting that active rDNA genes are hypomethylated. Direct evidence came from experiments with *Triticum* (Amado et al. 1997) and *Arabidopsis* (Lawrence et al. 2004), in which

DNA methylation was blocked by 5-aza-2'-deoxycytosine (5-aza) treatment leading to a hypomethylation state of DNA. In both cases a hybrid situation was investigated, in which one parental set of NORs was silenced due to nucleolar dominance. The reduction of DNA methylation resulted in the reactivation of the silent loci, indicating that the mechanism of rDNA inactivation involves DNA methylation. The lab of Pikaard further demonstrated with Chromatin Immuno Precipitation (ChIP) that RNA polymerase I-associated sequences correspond to hypomethylated DNA and histone H3K4 methylation, an epigenetic mark for euchromatin, whereas the inactive rDNA genes showed hypermethylated DNA, histone H3K9 methylation and reduced histone acetylation (Lawrence et al. 2004). The authors concluded that DNA methylation and histone deacetylation are interdependent and integral to rRNA gene silencing. In another study they identified the histone deacetylase HDA6 as the key factor in rDNA silencing and nucleolar dominance (Earley et al. 2006). DNA methylation and histone deacetylation are therefore suggested to cooperate, forming a self-reinforcing cycle to repress rDNA activity.

3.4.3 Epigenetic Marks in the Nucleolus and Flanking NOR Chromocenters

The majority of rDNA genes are silent and form compact domains flanking the nucleolus. They are decorated with epigenetic marks for silencing such as DNA methylation and methylated H3K9, but also H3K27me (Table 3). In contrast rDNA is actively transcribed in the nucleolus. It is therefore remarkable that immunodetection of several isoforms of acetylated histone H4 (H4K5, H4K8, H4K12, H4K16) do not label the nucleolus in *Arabidopsis* leaves, flower buds, or root tip at any cell-cycle stage (Jasencakova et al. 2003; Probst et al. 2003, 2004; Soppe et al. 2002). An increase in histone acetylation resulted in a uniform distribution in the nucleolus (Probst et al. 2004). However, in the DNA methylation mutants *ddm1* and *met1*, the nucleolus remained unlabeled (Soppe et al. 2002). The only histone acetylation in wild-type plants was histone H3K9/18, which was present all over the root tip nucleus, including the nucleolus, although in a cell cycle-dependent manner. These data are in contrast to plants with large genomes, such as *Vicia faba*, barley, or onion (Jasencakova et al. 2000, 2001; Mayr et al. 2003), where several isoforms of acetylated histones were observed in the nucleolus of root cells during most cell-cycle stages. The difference between these plants and *Arabidopsis* may be because certain forms of histone acetylation have a distinct function in different species. For example, the H3K16 acetylation pattern in the nucleolus is different in *V. faba*, barley, and *Arabidopsis* (Fuchs et al. 2006). Alternatively, it is possible that the level of acetylated histone in the nucleolus of *Arabidopsis* is below the detection limit. Indeed, immunolabeling with antiserum against total core histones and against histone H1 shows very faint signals in the nucleolus (Figs. 3 and 4).

The rDNA genes are a target for silencing by HDA6 and DNA methylation (Probst et al. 2004). *Arabidopsis hda6* mutants display decondensation of NOR chromocenters, hyperacetylated histone H4, hypermethylated H3K4, and reduced rDNA methylation. The same role for HDA6 and DNA methylation is demonstrated

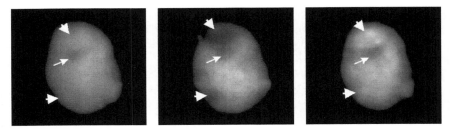

Fig. 3 Immunolabeling of acetylated H4K12 (*middle*) and total histone (*right*) in DAPI-stained nucleus (*left*) of *Arabidopsis*

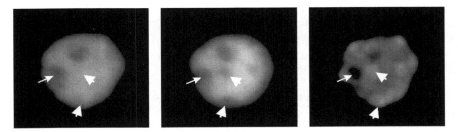

Fig. 4 Immunolabeling of methylated H3K4 (*middle*) and histone H1 (*right*) in DAPI-stained nucleus (*left*) of *Arabidopsis*. *Arrow*, nucleolus; *arrowhead*, chromocenter

in nucleolar dominance in the natural hybrid *A. suecica* (Earley et al. 2006; Pontes and Pikaard 2008). The presence of methylated cytosine attracts MBD proteins to chromocenters. However, AtMBD5 and AtMBD6 are preferentially located at NOR chromocenters, whereas AtMBD7 is observed in all chromocenters (Zemach et al. 2008). The authors propose that the single MBD motif in AtMBD5 and AtMBD6 is sufficient for localization to NOR chromocenters, while the triple MBD motif in AtMBD7 directs the protein to all chromocenters.

4 Concluding Remarks

The enormous potency of epigenetic combinations via modifications of DNA and histones, in conjunction with the combinations of code-reading domains and effector proteins likely provides sufficient variation to regulate the genetic information necessary to develop a eukaryotic organism. The nuclear architecture facilitates the optimal environment to establish the molecular interactions between epigenetic and genetic code. Chromosomes are folded into their territories in a way that corresponds to the linear arrangement of the DNA sequence creating domains such as chromocenters or nucleolus. Such domains concentrate chromatin regions that are targeted by the same chromatin modifiers. Chromosome folding is reversible, which allows chromatin to open and close to a certain extent depending on developmental stage or external signals. Plants and animals do not show large differences in epigenetic modules or chromatin organization. Variation in combining protein domains likely result in different interpretations of the epigenetic code, but the language is the same. In both kingdoms the assignment of the epigenetic code is to protect the genetic code (against external and internal mobile genetic elements) and to control the expression of the genetic information.

Our knowledge of the epigenetic code and the nuclear architecture has greatly increased during the past ten years. Yet the progress of high-throughput technologies has demonstrated that we are only at the beginning of another field to be explored. For example, proteome studies of the nucleolus revealed several hundreds of proteins that are not involved in ribosome biogenesis but in the modification of small RNAs, telomerase maturation, cell cycle and cell stress (Lo et al. 2006). Apparently, the nucleolus is an integrated domain for ribosome synthesis and many other functions. Indeed nuclear structures such as Cajal bodies and PML bodies associate with the nucleolus (Condemine et al. 2007; Li et al. 2006). Extensive chromatin immunoprecipitation studies have generated huge data sets revealing the epigenetic landscape along entire chromosomes (Turck et al. 2007; Zhang et al. 2007). If we can efficiently transform this vast growing body of data into knowledge, we may expect that within a few years we will be able to understand how epigenetic mechanisms control the nuclear organization and vice versa how the nuclear architecture regulates the establishment of the epigenetic code.

Acknowledgments The author wishes to acknowledge Z. Jasencakova for providing the image of the barley nucleus and H. de Jong and F. Tessadori for useful comments and discussions.

References

Agalioti T, Chen G, Thanos D (2002) Deciphering the transcriptional histone acetylation code for a human gene. Cell 111:381–392

Ahmad K, Henikoff S (2002) Histone H3 variants specify modes of chromatin assembly. Proc Natl Acad Sci USA 99(Suppl 4):16477–16484

Amado L, Abranches R, Neves N, Viegas W (1997) Development-dependent inheritance of 5-azacytidine-induced epimutations in triticale: analysis of rDNA expression patterns. Chromosome Res 5:445–450

Baumbusch LO, Thorstensen T, Krauss V, Fischer A, Naumann K, Assalkhou R, Schulz I, Reuter G, Aalen RB (2001) The *Arabidopsis thaliana* genome contains at least 29 active genes encoding SET domain proteins that can be assigned to four evolutionarily conserved classes. Nucleic Acids Res 29:4319–4333

Bender J (2004) DNA methylation and epigenetics. Annu Rev Plant Biol 55:41–68

Bennett MD, Leitch IJ, Price HJ, Johnston JS (2003) Comparisons with *Caenorhabditis* (approximately 100 Mb) and *Drosophila* (approximately 175 Mb) using flow cytometry show genome size in Arabidopsis to be approximately 157 Mb and thus approximately 25% larger than the

Arabidopsis genome initiative estimate of approximately 125 Mb. Ann Bot (Lond) 91:547–557

Berg A, Meza TJ, Mahic M, Thorstensen T, Kristiansen K, Aalen RB (2003) Ten members of the Arabidopsis gene family encoding methyl-CpG-binding domain proteins are transcriptionally active and at least one, AtMBD11, is crucial for normal development. Nucleic Acids Res 31:5291–5304

Berger SL (2007) The complex language of chromatin regulation during transcription. Nature 447:407–412

Bergmuller E, Gehrig PM, Gruissem W (2007) Characterization of post-translational modifications of histone H2B-variants isolated from *Arabidopsis thaliana*. J Proteome Res 6:3655–3668

Bernstein E, Duncan EM, Masui O, Gil J, Heard E, Allis CD (2006) Mouse polycomb proteins bind differentially to methylated histone H3 and RNA and are enriched in facultative heterochromatin. Mol Cell Biol 26:2560–2569

Blower MD, Sullivan BA, Karpen GH (2002) Conserved organization of centromeric chromatin in flies and humans. Dev Cell 2:319–330

Boisvert FM, van Koningsbruggen S, Navascues J, Lamond AI (2007) The multifunctional nucleolus. Nat Rev Mol Cell Biol 8:574–585

Brown SW (1966) Heterochromatin. Science 151:417–425

Bruce K, Myers FA, Mantouvalou E, Lefevre P, Greaves I, Bonifer C, Tremethick DJ, Thorne AW, Crane-Robinson C (2005) The replacement histone H2A.Z in a hyperacetylated form is a feature of active genes in the chicken. Nucleic Acids Res 33:5633–5639

Ceccarelli M, Morosi L, Cionini PG (1998) Chromocenter association in plant cell nuclei: determinants, functional significance, and evolutionary implications. Genome 41:96–103

Cloix C, Tutois S, Mathieu O, Cuvillier C, Espagnol MC, Picard G, Tourmente S (2000) Analysis of 5S rDNA arrays in *Arabidopsis thaliana*: physical mapping and chromosome-specific polymorphisms. Genome Res 10:679–690

Comings DE (1980) Arrangement of chromatin in the nucleus. Hum Genet 53:131–143

Condemine W, Takahashi Y, Le Bras M, de The H (2007) A nucleolar targeting signal in PML-I addresses PML to nucleolar caps in stressed or senescent cells. J Cell Sci 120:3219–3227

Copenhaver GP, Pikaard CS (1996) Two-dimensional RFLP analyses reveal megabase-sized clusters of rRNA gene variants in *Arabidopsis thaliana*, suggesting local spreading of variants as the mode for gene homogenization during concerted evolution. Plant J 9:273–282

Cremer T, Cremer C (2001) Chromosome territories, nuclear architecture and gene regulation in mammalian cells. Nat Rev Genet 2:292–301

de Jong JH, Fransz P, Zabel P (1999) High resolution FISH in plants – techniques and applications. Trends Plant Sci 4:258–263

Dorigo B, Schalch T, Kulangara A, Duda S, Schroeder RR, Richmond TJ (2004) Nucleosome arrays reveal the two-start organization of the chromatin fiber. Science 306:1571–1573

Earley K, Lawrence RJ, Pontes O, Reuther R, Enciso AJ, Silva M, Neves N, Gross M, Viegas W, Pikaard CS (2006) Erasure of histone acetylation by Arabidopsis HDA6 mediates large-scale gene silencing in nucleolar dominance. Genes Dev 20:1283–1293

Finch JT, Klug A (1976) Solenoidal model for superstructure in chromatin. Proc Natl Acad Sci USA 73:1897–1901

Finnegan EJ, Dennis ES (2007) Vernalization-induced trimethylation of histone H3 lysine 27 at FLC is not maintained in mitotically quiescent cells. Curr Biol 17:1978–1983

Fischle W, Wang Y, Jacobs SA, Kim Y, Allis CD, Khorasanizadeh S (2003) Molecular basis for the discrimination of repressive methyl-lysine marks in histone H3 by Polycomb and HP1 chromodomains. Genes Dev 17:1870–1881

Flavell RB (1986) Repetitive DNA and chromosome evolution in plants. Philos Trans R Soc Lond B Biol Sci 312:227–242

Fransz P, Armstrong S, Alonso-Blanco C, Fischer TC, Torres-Ruiz RA, Jones G (1998) Cytogenetics for the model system Arabidopsis thaliana. Plant J 13:867–876

Fransz PF, Armstrong S, de Jong JH, Parnell LD, van Drunen C, Dean C, Zabel P, Bisseling T, Jones GH (2000) Integrated cytogenetic map of chromosome arm 4S of *A*. Thaliana: structural organization of heterochromatic knob and centromere region. Cell 100:367–376

Fransz P, De Jong JH, Lysak M, Castiglione MR, Schubert I (2002) Interphase chromosomes in Arabidopsis are organized as well defined chromocenters from which euchromatin loops emanate. Proc Natl Acad Sci USA 99:14584–14589

Fransz P, ten Hoopen R, Tessadori F (2006) Composition and formation of heterochromatin in Arabidopsis thaliana. Chromosome Res 14:71–82

Fuchs J, Demidov D, Houben A, Schubert I (2006) Chromosomal histone modification patterns – from conservation to diversity. Trends Plant Sci 11:199–208

Garcia BA, Shabanowitz J, Hunt DF (2007) Characterization of histones and their post-translational modifications by mass spectrometry. Curr Opin Chem Biol 11:66–73

Grummt I (2003) Life on a planet of its own: regulation of RNA polymerase I transcription in the nucleolus. Genes Dev 17:1691–1702

Guenatri M, Bailly D, Maison C, Almouzni G (2004) Mouse centric and pericentric satellite repeats form distinct functional heterochromatin. J Cell Biol 166:493–505

Hassan AH, Awad S, Al-Natour Z, Othman S, Mustafa F, Rizvi TA (2007) Selective recognition of acetylated histones by bromodomains in transcriptional co-activators. Biochem J 402:125–133

Houben A, Demidov D, Gernand D, Meister A, Leach CR, Schubert I (2003) Methylation of histone H3 in euchromatin of plant chromosomes depends on basic nuclear DNA content. Plant J 33:967–973

Human Genome Sequencing C (2004) Finishing the euchromatic sequence of the human genome. Nature 431:931–945

Jackson JP, Lindroth AM, Cao X, Jacobsen SE (2002) Control of CpNpG DNA methylation by the KRYPTONITE histone H3 methyltransferase. Nature 416:556–560

Jacob ST, Ghosh AK (1999) Control of RNA polymerase I-directed transcription: recent trends. J Cell Biochem 75(S32):41–50

Jasencakova Z, Meister A, Walter J, Turner BM, Schubert I (2000) Histone H4 acetylation of euchromatin and heterochromatin is cell cycle dependent and correlated with replication rather than with transcription. Plant Cell 12:2087–2100

Jasencakova Z, Meister A, Schubert I (2001) Chromatin organization and its relation to replication and histone acetylation during the cell cycle in barley. Chromosoma 110:83–92

Jasencakova Z, Soppe WJ, Meister A, Gernand D, Turner BM, Schubert I (2003) Histone modifications in Arabidopsis – high methylation of H3 lysine 9 is dispensable for constitutive heterochromatin. Plant J 33:471–480

Jimenez-Garcia LF, Fragoso-Soriano R (2000) Atomic force microscopy of the cell nucleus. J Struct Biol 129:218–222

Johnson L, Mollah S, Garcia BA, Muratore TL, Shabanowitz J, Hunt DF, Jacobsen SE (2004) Mass spectrometry analysis of Arabidopsis histone H3 reveals distinct combinations of post-translational modifications. Nucleic Acids Res 32:6511–6518

Johnson LM, Bostick M, Zhang X, Kraft E, Henderson I, Callis J, Jacobsen SE (2007) The SRA methyl-cytosine-binding domain links DNA and histone methylation. Curr Biol 17:379–384

Kotake T, Takada S, Nakahigashi K, Ohto M, Goto K (2003) Arabidopsis TERMINAL FLOWER 2 gene encodes a heterochromatin protein 1 homolog and represses both FLOWERING LOCUS T to regulate flowering time and several floral homeotic genes. Plant Cell Physiol 44:555–564

Kulikova O, Gualtieri G, Geurts R, Kim DJ, Cook D, Huguet T, de Jong JH, Fransz PF, Bisseling T (2001) Integration of the FISH pachytene and genetic maps of Medicago truncatula. Plant J 27:49–58

Laibach F (1907) Zur Frage nach der Individualitat der Chromosomen im Pflanzenreich. Beiheft zum Bot. Centralblatt 22:191–210

Lawrence RJ, Earley K, Pontes O, Silva M, Chen ZJ, Neves N, Viegas W, Pikaard CS (2004) A concerted DNA methylation/histone methylation switch regulates rRNA gene dosage control and nucleolar dominance. Mol Cell 13:599–609

Lehnertz B, Ueda Y, Derijck AA, Braunschweig U, Perez-Burgos L, Kubicek S, Chen T, Li E, Jenuwein T, Peters AH (2003) Suv39h-mediated histone H3 lysine 9 methylation directs DNA methylation to major satellite repeats at pericentric heterochromatin. Curr Biol 13:1192–1200

Leuba SH, Yang G, Robert C, Samori B, van Holde K, Zlatanova J, Bustamante C (1994) Three-dimensional structure of extended chromatin fibers as revealed by tapping-mode scanning force microscopy. Proc Natl Acad Sci USA 91:11621–11625

Li A, Eirin-Lopez JM, Ausio J (2005) H2AX: tailoring histone H2A for chromatin-dependent genomic integrity. Biochem Cell Biol 83:505–515

Li CF, Pontes O, El-Shami M, Henderson IR, Bernatavichute YV, Chan SW, Lagrange T, Pikaard CS, Jacobsen SE (2006) An ARGONAUTE4-containing nuclear processing center colocalized with Cajal bodies in *Arabidopsis thaliana*. Cell 126:93–106

Libault M, Tessadori F, Germann S, Snijder B, Fransz P, Gaudin V (2005) The Arabidopsis LHP1 protein is a component of euchromatin. Planta 222:910–925

Lindroth AM, Shultis D, Jasencakova Z, Fuchs J, Johnson L, Schubert D, Patnaik D, Pradhan S, Goodrich J, Schubert I, Jenuwein T, Khorasanizadeh S, Jacobsen SE (2004) Dual histone H3 methylation marks at lysines 9 and 27 required for interaction with CHROMOMETHYLASE3. EMBO J 23:4286–4296

Liu S, Yu Y, Ruan Y, Meyer D, Wolff M, Xu L, Wang N, Steinmetz A, Shen WH (2007) Plant SET- and RING-associated domain proteins in heterochromatinization. Plant J 52:914–926

Lo SJ, Lee CC, Lai HJ (2006) The nucleolus: reviewing oldies to have new understandings. Cell Res 16:530–538

Lorincz MC, Dickerson DR, Schmitt M, Groudine M (2004) Intragenic DNA methylation alters chromatin structure and elongation efficiency in mammalian cells. Nat Struct Mol Biol 11:1068–1075

Luger K, Mader AW, Richmond RK, Sargent DF, Richmond TJ (1997) Crystal structure of the nucleosome core particle at 2.8 Å resolution. Nature 389:251–260

Maluszynska J, Heslop-Harrison JS (1991) Localization of tandemly repeated DNA sequences in *Arabidopsis thaliana*. Plant J 1:159–166

Manuelidis L (1990) A view of interphase chromosomes. Science 250:1533–1540

Margueron R, Trojer P, Reinberg D (2005) The key to development: interpreting the histone code? Curr Opin Genet Dev 15:163–176

Martens JH, O'Sullivan RJ, Braunschweig U, Opravil S, Radolf M, Steinlein P, Jenuwein T (2005) The profile of repeat-associated histone lysine methylation states in the mouse epigenome. EMBO J 24:800–812

Mathieu O, Probst AV, Paszkowski J (2005) Distinct regulation of histone H3 methylation at lysines 27 and 9 by CpG methylation in Arabidopsis. EMBO J 24:2783–2791

Mayr C, Jasencakova Z, Meister A, Schubert I, Zink D (2003) Comparative analysis of the functional genome architecture of animal and plant cell nuclei. Chromosome Res 11:471–484

McKittrick E, Gafken PR, Ahmad K, Henikoff S (2004) Histone H3.3 is enriched in covalent modifications associated with active chromatin. Proc Natl Acad Sci USA 101:1525–1530

Meneghini MD, Wu M, Madhani HD (2003) Conserved histone variant H2A.Z protects euchromatin from the ectopic spread of silent heterochromatin. Cell 112:725–736

Montijn MB, ten Hoopen R, Fransz PF, Oud JL, Nanninga N (1998) Characterisation of the nucleolar organising regions during the cell cycle in two varieties of *Petunia hybrida* as visualised by fluorescence in situ hybridisation and silver staining. Chromosoma 107:80–86

Muscarella DE, Vogt VM, Bloom SE (1985) The ribosomal RNA gene cluster in aneuploid chickens: evidence for increased gene dosage and regulation of gene expression. J Cell Biol 101:1749–1756

Mylne JS, Barrett L, Tessadori F, Mesnage S, Johnson L, Bernatavichute YV, Jacobsen SE, Fransz P, Dean C (2006) LHP1, the Arabidopsis homologue of HETEROCHROMATIN PROTEIN1, is required for epigenetic silencing of FLC. Proc Natl Acad Sci USA 103:5012–5017

Nakahigashi K, Jasencakova Z, Schubert I, Goto K (2005) The Arabidopsis heterochromatin protein1 homolog (TERMINAL FLOWER2) silences genes within the euchromatic region but not genes positioned in heterochromatin. Plant Cell Physiol 46:1747–1756

Naumann K, Fischer A, Hofmann I, Krauss V, Phalke S, Irmler K, Hause G, Aurich AC, Dorn R, Jenuwein T, Reuter G (2005) Pivotal role of AtSUVH2 in heterochromatic histone methylation and gene silencing in Arabidopsis. EMBO J 24:1418–1429

Ohmido N, Kijima K, Ashikawa I, de Jong JH, Fukui K (2001) Visualization of the terminal structure of rice chromosomes 6 and 12 with multicolor FISH to chromosomes and extended DNA fibers. Plant Mol Biol 47:413–421

Okitsu CY, Hsieh CL (2007) DNA methylation dictates histone H3K4 methylation. Mol Cell Biol 27:2746–2757

Oudet P, Gross-Bellard M, Chambon P (1975) Electron microscopic and biochemical evidence that chromatin structure is a repeating unit. Cell 4:281–300

Pecinka A, Schubert V, Meister A, Kreth G, Klatte M, Lysak MA, Fuchs J, Schubert I (2004) Chromosome territory arrangement and homologous pairing in nuclei of Arabidopsis thaliana are predominantly random except for NOR-bearing chromosomes. Chromosoma 113:258–269

Pontes O, Pikaard CS (2008) siRNA and miRNA processing: new functions for Cajal bodies. Curr Opin Genet Dev 18:197–203

Preuss S, Pikaard CS (2007) rRNA gene silencing and nucleolar dominance: insights into a chromosome-scale epigenetic on/off switch. Biochim Biophys Acta 1769:383–392

Probst AV, Fransz PF, Paszkowski J, Mittelsten Scheid O (2003) Two means of transcriptional reactivation within heterochromatin. Plant J 33:743–749

Probst AV, Fagard M, Proux F, Mourrain P, Boutet S, Earley K, Lawrence RJ, Pikaard CS, Murfett J, Furner I, Vaucheret H, Mittelsten Scheid O (2004) Arabidopsis histone deacetylase HDA6 is required for maintenance of transcriptional gene silencing and determines nuclear organization of rDNA repeats. Plant Cell 16:1021–1034

Ramsahoye BH, Biniszkiewicz D, Lyko F, Clark V, Bird AP, Jaenisch R (2000) Non-CpG methylation is prevalent in embryonic stem cells and may be mediated by DNA methyltransferase 3a. Proc Natl Acad Sci USA 97:5237–5242

Rangasamy D, Berven L, Ridgway P, Tremethick DJ (2003) Pericentric heterochromatin becomes enriched with H2A.Z during early mammalian development. EMBO J 22:1599–1607

Rea S, Eisenhaber F, O'Carroll D, Strahl BD, Sun ZW, Schmid M, Opravil S, Mechtler K, Ponting CP, Allis CD, Jenuwein T (2000) Regulation of chromatin structure by site-specific histone H3 methyltransferases. Nature 406:593–599

Robinson PJ, Fairall L, Huynh VA, Rhodes D (2006) EM measurements define the dimensions of the "30-nm" chromatin fiber: evidence for a compact, interdigitated structure. Proc Natl Acad Sci USA 103:6506–6511

Rogers SO, Bendich AJ (1987) Heritability and variability in ribosomal RNA genes of Vicia faba. Genetics/Society 117:285–295

Santos AP, Abranches R, Stoger E, Beven A, Viegas W, Shaw PJ (2002) The architecture of interphase chromosomes and gene positioning are altered by changes in DNA methylation and histone acetylation. J Cell Sci 115:4597–4605

Scebba F, Bernacchia G, De Bastiani M, Evangelista M, Cantoni RM, Cella R, Locci MT, Pitto L (2003) Arabidopsis MBD proteins show different binding specificities and nuclear localization. Plant Mol Biol 53:715–731

Schalch T, Duda S, Sargent DF, Richmond TJ (2005) X-ray structure of a tetranucleosome and its implications for the chromatin fibre. Nature 436:138–141

Schotta G, Lachner M, Sarma K, Ebert A, Sengupta R, Reuter G, Reinberg D, Jenuwein T (2004) A silencing pathway to induce H3-K9 and H4-K20 trimethylation at constitutive heterochromatin. Genes Dev 18:1251–1262

Seet BT, Dikic I, Zhou MM, Pawson T (2006) Reading protein modifications with interaction domains. Nat Rev Mol Cell Biol 7:473–483

Shogren-Knaak M, Ishii H, Sun JM, Pazin MJ, Davie JR and Peterson CL (2006) Histone H4-K16 acetylation controls chromatin structure and protein interactions. Science 311:844–847

Soppe WJ, Jasencakova Z, Houben A, Kakutani T, Meister A, Huang MS, Jacobsen SE, Schubert I, Fransz PF (2002) DNA methylation controls histone H3 lysine 9 methylation and heterochromatin assembly in Arabidopsis. EMBO J 21:6549–6559

Springer NM, Kaeppler SM (2005) Evolutionary divergence of monocot and dicot methyl-CpG-binding domain proteins. Plant Physiol 138:92–104

Strahl BD, Allis CD (2000) The language of covalent histone modifications. Nature 403:41–45

Tamaru H, Selker EU (2001) A histone H3 methyltransferase controls DNA methylation in Neurospora crassa. Nature 414:277–283

Tessadori F, Chupeau MC, Chupeau Y, Knip M, Germann S, van Driel R, Fransz P, Gaudin V (2007a) Large-scale dissociation and sequential reassembly of pericentric heterochromatin in dedifferentiated Arabidopsis cells. J Cell Sci 120:1200–1208

Tessadori F, Schulkes RK, van Driel R, Fransz P (2007b) Light-regulated large-scale reorganization of chromatin during the floral transition in Arabidopsis. Plant J 50:848–857

The Arabidopsis Genome Initiative (2000) Analysis of the genome sequence of the flowering plant Arabidopsis thaliana. Nature 408:796–815

Thoma F, Koller T, Klug A (1979) Involvement of histone H1 in the organization of the nucleosome and of the salt-dependent superstructures of chromatin. J Cell Biol 83:403–427

Tschochner H, Hurt E (2003) Pre-ribosomes on the road from the nucleolus to the cytoplasm. Trends Cell Biol 13:255–263

Turck F, Roudier F, Farrona S, Martin-Magniette ML, Guillaume E, Buisine N, Gagnot S, Martienssen RA, Coupland G, Colot V (2007) Arabidopsis TFL2/LHP1 specifically associates with genes marked by trimethylation of histone H3 lysine 27. PLoS Genet 3:e86

Turner BM (2000) Histone acetylation and an epigenetic code. Bioessays 22:836–845

van Driel R, Fransz PF, Verschure PJ (2003) The eukaryotic genome: a system regulated at different hierarchical levels. J Cell Sci 116:4067–4075

Verschure PJ, van der Kraan I, Manders EM, Hoogstraten D, Houtsmuller AB, van Driel R (2003) Condensed chromatin domains in the mammalian nucleus are accessible to large macromolecules. EMBO Rep 4:861–866

Waterston RH, Lindblad-Toh K, Birney E, Rogers J, Abril JF, Agarwal P, Agarwala R, Ainscough R, Alexandersson M, An P, Antonarakis SE, Attwood J, Baertsch R, Bailey J, Barlow K, Beck S, Berry E, Birren B, Bloom T, Bork P, Botcherby M, Bray N, Brent MR, Brown DG, Brown SD, Bult C, Burton J, Butler J, Campbell RD, Carninci P, Cawley S, Chiaromonte F, Chinwalla AT, Church DM, Clamp M, Clee C, Collins FS, Cook LL, Copley RR, Coulson A, Couronne O, Cuff J, Curwen V, Cutts T, Daly M, David R, Davies J, Delehaunty KD, Deri J, Dermitzakis ET, Dewey C, Dickens NJ, Diekhans M, Dodge S, Dubchak I, Dunn DM, Eddy SR, Elnitski L, Emes RD, Eswara P, Eyras E, Felsenfeld A, Fewell GA, Flicek P, Foley K, Frankel WN, Fulton LA, Fulton RS, Furey TS, Gage D, Gibbs RA, Glusman G, Gnerre S, Goldman N, Goodstadt L, Grafham D, Graves TA, Green ED, Gregory S, Guigo R, Guyer M, Hardison RC, Haussler D, Hayashizaki Y, Hillier LW, Hinrichs A, Hlavina W, Holzer T, Hsu F, Hua A, Hubbard T, Hunt A, Jackson I, Jaffe DB, Johnson LS, Jones M, Jones TA, Joy A, Kamal M, Karlsson EK, Karolchik D, Kasprzyk A, Kawai J, Keibler E, Kells C, Kent WJ, Kirby A, Kolbe DL, Korf I, Kucherlapati RS, Kulbokas EJ, Kulp D, Landers T, Leger JP, Leonard S, Letunic I, Levine R, Li J, Li M, Lloyd C, Lucas S, Ma B, Maglott DR, Mardis ER, Matthews L, Mauceli E, Mayer JH, McCarthy M, McCombie WR, McLaren S, McLay K, McPherson JD, Meldrim J, Meredith B, Mesirov JP, Miller W, Miner TL, Mongin E, Montgomery KT, Morgan M, Mott R, Mullikin JC, Muzny DM, Nash WE, Nelson JO, Nhan MN, Nicol R, Ning Z, Nusbaum C, O'Connor MJ, Okazaki Y, Oliver K, Overton-Larty E, Pachter L, Parra G, Pepin KH, Peterson J, Pevzner P, Plumb R, Pohl CS, Poliakov A, Ponce TC, Ponting CP, Potter S, Quail M, Reymond A, Roe BA, Roskin KM, Rubin EM, Rust AG, Santos R, Sapojnikov V, Schultz B, Schultz J, Schwartz MS, Schwartz S, Scott C, Seaman S, Searle S, Sharpe T, Sheridan A, Shownkeen R, Sims S, Singer JB, Slater G, Smit A, Smith DR, Spencer B, Stabenau A, Stange-Thomann N, Sugnet C, Suyama M, Tesler G,

Thompson J, Torrents D, Trevaskis E, Tromp J, Ucla C, Ureta-Vidal A, Vinson JP, Von Niederhausern AC, Wade CM, Wall M, Weber RJ, Weiss RB, Wendl MC, West AP, Wetterstrand K, Wheeler R, Whelan S, Wierzbowski J, Willey D, Williams S, Wilson RK, Winter E, Worley KC, Wyman D, Yang S, Yang SP, Zdobnov EM, Zody MC, Lander ES (2002) Initial sequencing and comparative analysis of the mouse genome. Nature 420:520–562

Woo HR, Pontes O, Pikaard CS, Richards EJ (2007) VIM1, a methylcytosine-binding protein required for centromeric heterochromatinization. Genes Dev 21:267–277

Wu C, Bassett A, Travers A (2007) A variable topology for the 30-nm chromatin fibre. EMBO Rep 8:1129–1134

Zemach A, Grafi G (2003) Characterization of *Arabidopsis thaliana* methyl-CpG-binding domain (MBD) proteins. Plant J 34:565–572

Zemach A, Li Y, Wayburn B, Ben-Meir H, Kiss V, Avivi Y, Kalchenko V, Jacobsen SE, Grafi G (2005) DDM1 binds Arabidopsis methyl-CpG binding domain proteins and affects their sub-nuclear localization. Plant Cell 17:1549–1558

Zemach A, Gaspan O, Grafi G (2008) The three methyl-CpG-binding domains of AtMBD7 control its subnuclear localization and mobility. J Biol Chem 283:8406–8411

Zhang X, Germann S, Blus BJ, Khorasanizadeh S, Gaudin V, Jacobsen SE (2007) The Arabidopsis LHP1 protein colocalizes with histone H3 Lys27 trimethylation. Nat Struct Mol Biol 14:869–871

Zhang WZ, Lee HR, Koo DH, Jiang J (2008) Epigenetic modification of centromeric chromatin: hypomethylation of DNA sequences in the CENH3-associated chromatin in Arabidopsis thaliana and maize. Plant Cell 20:25–34

Zhong XB, Fransz PF, Wennekes-Eden J, Ramanna MS, van Kammen A, Zabel P, Hans de Jong J (1998) FISH studies reveal the molecular and chromosomal organization of individual telomere domains in tomato. Plant J 13:507–517

Integration of *Agrobacterium* T-DNA in Plant Cells

Mery Dafny-Yelin, Andriy Tovkach, and Tzvi Tzfira (✉)

Abstract *Agrobacterium*-mediated genetic transformation is a process by which the bacterium delivers a specific DNA molecule into plant cells. The transferred DNA molecule (T-DNA) stably integrates into the host genome and is expressed there. *Agrobacterium*-mediated genetic transformation is widely used for the production of transgenic plants useful for basic plant research and biotechnology, yet the mechanisms by which the T-DNA integrates into the host genome are still poorly understood. Furthermore, we have only recently begun to reveal the important functions of plant factors in the integration process. In this chapter, we describe the current knowledge on the bacterial and host factors and the cellular mechanisms that govern the integration of T-DNA molecules into plant cells. We follow the long line of genetic, functional, and biochemical studies which have paved the way for establishing the different integration models, and we describe possibilities for controlling the T-DNA integration process in order to achieve the most desirable gene-targeting technology for plant species.

Abbreviations ds: double-stranded; DSB: double-strand breaks; GT: gene targeting; HR: homologous recombination; NHEJ: non homologous end joining; ss: single-stranded; T-DNA: transferred DNA; T-strand: the single-stranded DNA form of the T-DNA; ZFNs: zinc-finger nucleases

1 Introduction

Plant genetic transformation and the production of transgenic plants are essential for modern plant research and biotechnology. The most common vector used today for the genetic transformation of various model and crop plants is *Agrobacterium*

T. Tzfira
Department of Molecular, Cellular and Developmental Biology, The University of Michigan, Ann Arbor, MI 48109 USA
e-mail: ttzfira@umich.edu

Plant Cell Monogr, doi:10.1007/7089_2008_28
© Springer-Verlag Berlin Heidelberg 2008

tumefaciens (reviewed in Tzfira and Citovsky 2006). This soil-borne bacterium is capable of transforming its host by delivering a well-defined fraction of its own genome as a single-stranded (ss) DNA molecule into the host cell. This ssDNA molecule, designated transferred DNA (T-DNA), ultimately integrates into the host genome and is expressed there (for recent reviews see Citovsky et al. 2007; Gelvin 2003; McCullen and Binns 2006; Tzfira and Citovsky 2006). The natural host range of wild-type *Agrobacterium* is limited to certain dicotyledonous plants, where it is considered the causative agent of "crown-gall" disease (de Cleene and de Ley 1976; Otten et al. 2008). Nevertheless, the current host spectrum of *Agrobacterium* spans a much wider range of species, as the bacteria have been reported to transform, at least under controlled laboratory conditions, not only an ever-increasing number of plant species (reviewed in Banta and Montenegro 2008), but also eukaryotes from other families, ranging from lower eukaryotes such as yeast (Bundock et al. 1995; Piers et al. 1996) and other fungi (e.g., de Groot et al. 1998; Godio et al. 2004; Michielse et al. 2005) to higher eukaryotes, such as human cells (Kunik et al. 2001) (reviewed in Lacroix et al. 2006b; Soltani et al. 2008). For several decades, *Agrobacterium* research has provided a fertile ground for various research groups interested not only in harnessing the bacterium's unique biology for the genetic transformation of various plant species (reviewed in Banta and Montenegro 2008; Gelvin 1998, 2003), but also in understanding the biological systems that are involved in the infection process (reviewed in Citovsky et al. 2007; Gelvin 2003; Tzfira and Citovsky 2002). Nevertheless, while *Agrobacterium* can be used in practice for transforming a wide range of scientifically and economically important species, many of the biological mechanisms that govern the transformation process are still poorly understood. Thus, for example, we know only little about the machineries and mechanisms that allow the transport of T-DNA molecules from the bacterium into the plant cells, and even less about the apparati and means governing the transport of T-DNA molecules through the host-cell cytoplasm toward the nucleus. More importantly, our understanding of the precise mechanisms controlling T-DNA integration into the host-cell genome is still lacking. Thus, we are currently very limited in our ability to influence the integration process and to control the outcome of the transformation event. In this chapter, we focus our discussion on the proteins that participate in the integration process (Table 1) and on the possible mechanisms that lead to integration of T-DNA molecules into the plant genome. We also discuss the possibility of controlling the integration process as a tool for achieving site-specific integration and gene targeting (GT) in plant species.

2 The Genetic Transformation Process: An Overview

The *Agrobacterium*-mediated genetic transformation process has been the subject of numerous reviews and the reader is referred to them for in-depth discussions (e.g., Citovsky et al. 2007; Gelvin 2003; McCullen and Binns 2006; Tzfira and Citovsky

Table 1 The function of bacterial and host proteins in the T-DNA integration process

Gene/protein/ mutant/ complex	Biological functions	Possible roles in T-DNA integration	Selected refs.
Bacterial proteins			
VirD2	Production of T-strand in the bacteria, leading T-DNA to the host cytoplasm, nuclear import of the T-complex	Required for precision of integration, possibly by protecting T-strands on their route to integration, possess an indirect role in T-complex targeting to transcriptionally active sites by interactions with TBP and with CAK2Ms, recruiting plant ligase(s) to points of integration	Bako et al. 2003; Ballas and Citovsky 1997; Mysore et al. 1998; Pansegrau et al. 1993; Tinland et al. 1992; Wu 2002; Ziemienowicz et al. 2001
VirE2	A ssDNA-binding protein, packaging shaping and protecting the T-strand on its way to the nucleus, assist with nuclear T-strand import	Indirect role in T-complex targeting to the genome via interactions with host proteins VIP1 and VIP2	Abu-Arish et al. 2004; Anand et al. 2007; Citovsky et al. 1989; Citovsky et al. 1992; Howard and Citovsky 1990; Loyter et al. 2005; Ward and Zambryski 2001; Ziemienowicz et al. 2001
VirF	F-box bacterial proteins, function in stripping the T-complex from its escorting proteins VirE2 and VIP1	Indirect role by stripping the T-complex to naked T-strand	Schrammeijer et al. 2001; Tzfira et al. 2004b
Yeast proteins			
Ku70 mutant	Ku70 is a dsDNA-binding protein, binds and stabilizes dsDNA ends, key protein in the NHEJ	Essential for NHEJ-mediated T-DNA integration, in its absence, T-DNA integration occurs only via HR.	van Attikum et al. 2001; van Attikum and Hooykaas 2003
lig4, mre11, rad50, xrs2 and *sir4* mutants	NHEJ proteins	Required for T-DNA integration via NHEJ	van Attikum et al. 2001; van Attikum and Hooykaas 2003
Rad52 mutant	Rad52 is a ssDNA-binding protein, binds and stabilizes ssDNA ends, key proteins in HR	Essential for HR-mediated T-DNA integration, in its absence, T-DNA integration occurs only via NHEJ	van Attikum et al. 2001; van Attikum and Hooykaas 2003

(continued)

Table 1 (continued)

Gene/protein/ mutant/ complex	Biological functions	Possible roles in T-DNA integration	Selected refs.
Rad51 mutant	Rad51 is ssDNA-binding protein which interacts with RPA complex and RAD52, required for HR	Required for T-DNA integration via HR	van Attikum et al. 2001; van Attikum and Hooykaas 2003
Plant proteins			
Arabidopsis ecotype UV-1	Unknown factor	UB-1 ecotype is deficient in T-DNA integration using root-transformation assay	Mysore et al. 2000a; Nam et al. 1997
ASK1	Part of the plant SCF complex, involved in targeted proteolysis	Indirect role by stripping the T-complex to naked T-strand	Schrammeijer et al. 2001; Tzfira et al. 2004b
CAF-1/*fas1*-4 mutant	Chromatin assembly factor 1 (CAF-1) is involved in nucleosome assembly; *fas1-4* mutants exhibit increased frequency of somatic HR	Frequency of T-DNA integration is elevated in *fas1-4* mutant. CAF may function in inhibiting T-DNA integration by protecting the chromosomal DNA, HR increased dramatically (96-fold)	Endo et al. 2006; Kirik et al. 2006
CAK2Ms and TBP	CAK2Ms and TBP from alfalfa cells are part of the plant transcription machinery	VirD2 interacts with CAK2Ms and TBP and becomes phosphorylated by CAK2Ms kinase, may participate in T-complex targeting to transcriptionally active sites	Bako et al. 2003
CyPs	Molecular chaperone, may possess nuclease activity	Binds to VirD2, may function in T-complex targeting to chromatin, and may function in digestion of target DNA integration site	Bako et al. 2003; Deng et al. 1998; Montague et al. 1997
H2A, *rat5* mutant	H2A-1 is a core histone, may participate in DNA packaging	*rat5* mutant is deficient in *Agrobacterium* T-DNA integration, H2A-1 gene correlates with susceptibility to *Agrobacterium*, interacts with VIP1 and may function in T-complex targeting to chromatin	Li et al. 2005a; Loyter et al. 2005; Mysore et al. 2000b; Yi et al. 2002

H2A/H2B H3/H4	Core histone proteins	Can complement the *rat5* mutation when over-expressed, possible functional redundancy with H2A-1	Yi et al. 2006
INO80	Chromatin remodeling complex subunit that is specific to HR, INO80 is a positive regulator of HR	Efficiency of T-DNA integration was not affected in *ino80* mutant	Fritsch et al. 2004; van Attikum et al. 2004
ku80 mutant	Ku80 is a dsDNA-binding protein, binds and stabilizes dsDNA ends, key protein in the NHEJ	*ku80* mutant is blocked at T-DNA integration in root transformation, *ku80* mutant may or may not be blocked in T-DNA integration using flower-dip transformation, KU80 binds and recombines dsT-DNA molecules	Friesner and Britt 2003; Gallego et al. 2003; Li et al. 2005b
lig4 mutant	Arabidopsis DNA ligase IV	The role of AtLIG4 in T-DNA integration is still controversial: it is dispensable for root transformation and is either required or not for germ-line transformation	Friesner and Britt 2003; van Attikum et al. 2003
LIG1	Arabidopsis DNA ligase I	AtLIG1 interacts with VirD2, is capable of ligating T-DNAs to model chromosomal DNA in vitro	Wu 2002
uvh1 and *rad5* mutants	Mutants are hypersensitive to radiation	The mutant *rad5* is deficient in T-DNA integration while the deficiency of *uvh1* in T-DNA integrations is controversial	Chateau et al. 2000; Nam et al. 1998; Preuss et al. 1999
VIP1	Putative transcriptional factor, activates pathogenesis-related gene	Binds to VirE2 and facilitate the nuclear import of T-complex, interacts with host histones and VIP2 and function in T-complex targeting to chromatin, interacts with VirF and assists function in stripping the T-complex from VirE2	Anand et al. 2007; Dafny-Yelin et al. 2008; Djamei et al. 2007; Li et al. 2005a; Tzfira et al. 2001; Tzfira et al. 2002; Tzfira et al. 2004b; Ward et al. 2002
VIP2	Putative negative transcriptional regulator	Interacts with VirE2 and VIP1, might participate in intranuclear transport of VirE2 and T-complexes to chromatin and/or in T-DNA integration	Anand et al. 2007

2002, 2006). Briefly, the transformation process is controlled by a combination of chromosomally (*chv*) and plasmid-virulence (*vir*)-encoded proteins. The latter are present on the bacterium's large tumor-inducing (Ti) plasmid, which also contains the T-DNA region (Fig. 1). The transformation begins with host-cell recognition by the *Agrobacterium* VirA/VirG sensory machinery, which responds to various signals that are typically secreted by wounded plant tissues. Recognition is followed by activation of the *vir* machinery and attachment of the bacterium to the host cell. Through the combined action of the VirD2/D1 endonuclease, the bacterium releases a ssDNA (T-strand) copy of its T-DNA region which is then transported into the host-cell cytoplasm through a VirB/VirD4 type IV secretion system (reviewed by Atmakuri and Christie 2008; Christie et al. 2005). The T-strand is not exported as a naked DNA molecule, but as a protein–DNA complex (designated immature T-complex) in which a single VirD2 molecule is covalently attached to its 5′ end. The VirB/VirD4 transport apparatus is also capable of transferring a series of other Vir proteins (e.g., VirE2, VirE3, and VirF), which function within the host cell and further assist with the transformation process (for recent reviews see Citovsky et al. 2007; Gelvin 2003; McCullen and Binns 2006; Tzfira and Citovsky 2006). Once inside the host cytoplasm, the immature T-complex is thought to become covered with numerous copies of VirE2, producing the mature T-complex (simply referred to as the T-complex). VirE2's association with the T-strand provides it with a defined structure (Abu-Arish et al. 2004; Citovsky et al. 1997) and protects it from endonucleases (Citovsky et al. 1989). Both VirE2 and VirD2, through interactions with various host and bacterial proteins, are thought to assist with the nuclear import of the T-complex (e.g., Ballas and Citovsky 1997; Howard et al. 1992; Lacroix et al. 2005, 2006a, 2008; Tzfira et al. 2001; Ziemienowicz et al. 2001; Zupan et al. 1996). Once inside the nucleus, the T-complex is targeted to points of integration, stripped of its escorting proteins and integrated into the host-cell genome, as described in the next sections.

3 Components and Functions of the Integration Machinery

3.1 The Substrate Molecule to Be Integrated

In order to analyze the mechanism of the integration process, we need to define its substrate, i.e., the structure of the integrating T-DNA molecule. The T-DNA molecule undergoes several structural steps to its point of integration, beginning at the

can potentially be converted to double-stranded intermediates which may (iv) or may not (v) contain a VirD2 molecule attached to the T-strand's original 5' end. While all the different T-DNA structures may potentially be used as substrates by various integration machineries, direct experimental evidence only exists for the integration of double-stranded T-DNA molecules which most likely do not contain a VirD2 molecule (v). See text for further information

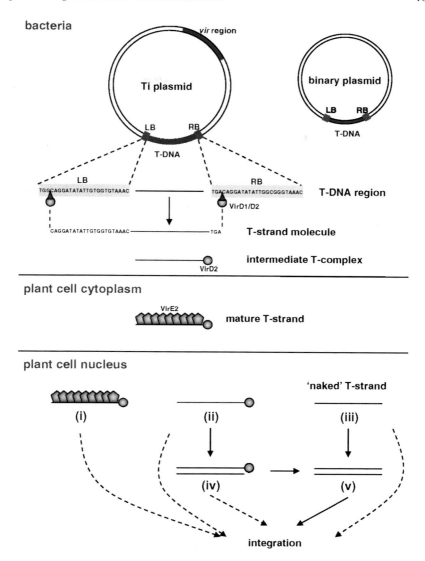

Fig. 1 *Agrobacterium* T-DNA: its various structures in the bacteria and plant cell compartments and its possible pathways for integration in the host nucleus. *Upper panel:* The T-DNA is defined, in the bacteria, by its left border (*LB*) and right border (*RB*). The T-DNA can reside on the Ti plasmid or a binary plasmid. A VirD2-VirD1 endonuclease complex, encoded by the bacterial *vir* region, is responsible for nicking between the third and fourth nucleotides on each border sequence and for the release of a mobile copy of the T-DNA. This mobile copy, designated intermediate T-complex, is composed of the T-strand and a single VirD2 molecule attached to its 5-end. *Center panel:* Within the plant-cell cytoplasm, the T-strand is thought to exist as a mature T-complex composed of the T-strand, a single VirD2 molecule and numerous VirE2 molecules which wrap, shape, and protect it on its way to the host-cell nucleus. *Bottom panel:* Several T-DNA substrates can potentially be used in the integration process, i.e. a mature T-complex (i), an partially stripped T-complex (ii) or a 'stripped/naked' T-strand (iii). The latter two T-DNA structures

bacterium where the T-DNA is defined by its borders, direct 25-bp repeats (Fig. 1) which are the only *cis* elements required for its excision from the Ti plasmids by the VirD2/D1 heterodimer (reviewed by Zambryski 1992). The T-DNA does not carry any specific sequences which might be required for its transport to the plant cell, nuclear import or integration. Thus, the entire native T-DNA region can be re-engineered and replaced by virtually any other DNA sequence of interest. Indeed, replacing the native T-DNA regions with various genes of interest is the molecular basis for the production of many genetically modified plants carrying agronomically and/or scientifically important genes (reviewed in Banta and Montenegro 2008). Furthermore, recombined T-DNA molecules need not reside on the bacterial Ti plasmid, and can be carried by small, autonomous binary plasmids (Fig. 1; Hellens et al. 2000). It should be noted, however, that it is likely that the mechanisms that govern the processing, release, transport, and integration of the T-strand from binary plasmids are similar to those that govern these processes for wild-type T-strands.

Inside the host-cell cytoplasm, the T-strand is assembled into a mature T-complex and guided into the host-cell nucleus (reviewed in Lacroix et al. 2006a, 2008; Tzfira et al. 2005). Thus, the initial substrate for the integration process is most likely to be the T-complex, rather than a naked T-strand molecule (Fig. 1). Indeed, the T-complex's escorting proteins have been reported to interact with host factors and to participate in various stages of the integration process, e.g., targeting the T-complex to points of integration (see further on). Nevertheless, the T-complex will eventually need to be stripped of its escorting proteins before or during its incorporation into the host-cell genome, thus leaving a partially or completely naked T-strand molecule (Fig. 1). Functional studies have revealed that T-strands can be complemented to double-stranded (ds) intermediates prior to their incorporation into the host genome (Chilton and Que 2003; Tzfira et al. 2003), but the integration of single-stranded molecules has never been ruled out. It is therefore likely that both ssT-DNA and dsT-DNA intermediates are substrates for the last step(s) of the integration process and that their incorporation into the host genome may be governed by different mechanisms, as we describe further on.

3.2   The Proteins Involved in T-DNA Integration

The bacterial T-strand-escorting proteins, VirE2 and VirD2, have both been suggested to participate in the integration process (Table 1). VirD2 has been suggested to act as a DNA ligase or integrase and is thus thought to directly function in the integration process by facilitating the incorporation of T-strand molecules into the host-cell genome (Pansegrau et al. 1993; Tinland et al. 1995). VirE2, on the other hand, has been proposed to function indirectly, most likely by protecting the T-strand from cellular endonucleases prior to its integration into the host-cell genome (Citovsky et al. 1989). More recent data, however, have revealed that VirD2 does not possess the biochemical activity needed for T-strand ligation into the host-cell genome (Ziemienowicz et al. 2000) and it is now accepted that both VirD2 and VirE2 act indirectly in the integration process by recruiting various plant factors and directing the T-strand to its points of integration (reviewed in Tzfira et al. 2004a; Ziemienowicz et al. 2008).

3.2.1 Functions of Bacterial Proteins in the Integration Process

VirD2, through its interaction with the T-strand's 5′ end, provides the T-strand with a direction and is thought to pilot it into the host-cell nucleus, through interactions with the plant's nuclear-import machinery (Ballas and Citovsky 1997). Because VirD2 functions as a DNA-processing protein in the bacteria, and because of the covalent nature of this interaction, it was also thought to function as a DNA-repair and processing protein during the integration process. It was initially suggested that VirD2 may function as an integrase and possibly a ligase in plant cells (Pansegrau et al. 1993; Tinland et al. 1995). Indeed, an H-R-Y motif, also found in bacteriophage λ integrase and other site-specific recombinases, has been identified in the VirD2 amino-acid sequence. More importantly, a mutation in that domain, which converted R to G, affected the precision of the T-DNA-integration process, as reflected by loss of conservation at the integrated T-DNA's right border (Tinland et al. 1995). The notion that VirD2 functions as a DNA ligase was supported by studies showing that VirD2 is not only capable of cleaving dsDNA molecules at border sequences, but also of ligating the resultant ssDNA molecule with another ssDNA molecule (Pansegrau et al. 1993). Interestingly, the mutation in H-R-Y did not affect the overall efficiency of the integration process (Tinland et al. 1995) and the in-vitro rejoining of VirD2-cleaved DNA substrate, in contrast to the random nature of the integration process, was sequence-specific (Pansegrau et al. 1993). These observations raised some doubt as to the functional role of VirD2 as a putative integrase and/or ligase, but did not preclude the possibility that VirD2 may still function as such during the integration process. However, using an in-vitro ligation assay, Ziemienowicz et al. (2000) revealed that VirD2 cannot ligate the 5′ end of T-DNA molecules to free 3′ ends of plant DNA. Because T-DNA ligation could be observed only when a commercial T4 DNA ligase or plant extracts from either tobacco BY-2 cells or pea axes were added to the ligation assay, the authors (Ziemienowicz et al. 2000) suggested that a plant DNA ligase(s), and not VirD2, is actually responsible for T-DNA ligation in plant cells. Recent observations that VirD2 can interact with the *Arabidopsis* type I DNA ligase suggest that the latter may function as the actual ligase of T-DNA molecules in plant cells (Wu 2002).

Similar to VirD2, VirE2 also binds to the T-strand and participates in various steps of the transformation process which precede the integration stage (Duckely and Hohn 2003; Ward and Zambryski 2001). One of the important roles of VirE2 is to protect the T-strand during its voyage through the cytoplasm and into the nucleus (reviewed in Lacroix et al. 2006a; Tzfira et al. 2005). Indeed, while two different structural models have been proposed to describe the cooperative binding of VirE2 to ssDNA, both models propose protection of the T-strand from degradation, possibly by cytoplasmic and nuclear endonucleases (Abu-Arish et al. 2004;

Citovsky et al. 1997; Volokhina and Chumakov 2007). That the genetic transformation of tobacco plants with a *virE2*-deficient *Agrobacterium* strain resulted in integration of truncated T-DNA molecules into the plant chromosomes (Rossi et al. 1996) further supports the notion that VirE2 protects the T-strand prior to its integration. VirE2 also plays an important role in the nuclear import and intranuclear transport of the T-complex in plant cells through interactions with various host and other bacterial factors (reviewed in Lacroix et al. 2006a; Tzfira et al. 2005), but is not likely to enzymatically participate in the integration process per se.

Although VirD2 and VirE2 may not have a direct role in the integration process, they both participate in targeting the T-DNA to points of integration via interactions with various host factors, or by recruiting plant enzymes that are involved in DNA repair or recombination, to the integration site. VirD2 has been shown to interact with CAK2Ms and TATA-box-binding protein (TBP) in the nuclei of alfalfa cells (Bako et al. 2003). CAK2Ms is a conserved plant orthologue of cyclin-dependent kinase-activating kinases and it functions by binding and phosphorylating the C-terminal regulatory domain of RNA polymerase II's largest subunit, which then recruits TBP to transcription sites. Furthermore, VirD2 was not only shown to interact with CAK2Ms and TBP, it was also reported to become phosphorylated by CAK2Ms kinase. These studies suggest that VirD2 may act by directing the T-complex to transcriptionally active genomic sites, which may function as "hot spots" for T-DNA integration. VirD2 was also shown to interact with several plant cyclophilins (CyPs), such as RocA, Roc4 and CypA (Bako et al. 2003; Deng et al. 1998). CyPs are conserved peptidyl-prolyl *cis-trans* isomerases that function as molecular chaperones. Inhibition of VirD2-CypA by cyclosporin A resulted in inhibition of *Agrobacterium*-mediated transformation of *Arabidopsis* and tobacco plants. It was thus suggested that CyPs may act as molecular chaperones that maintain VirD2's conformation throughout its journey to the nucleus. Since several CyPs have been shown to possess nuclease activity (e.g., Montague et al. 1997), it may also be that they function during the integration process. It should be noted that VirD2 was also capable of interacting with *Arabidopsis* type I DNA ligase (Wu 2002), which suggests that it may function by recruiting this DNA-repair protein to the integration site.

The ability of VirE2 to shape and protect the T-strand during its journey to the nucleus and guide it into the nucleus may be this bacterial protein's main functions (reviewed in Lacroix et al. 2006a; Tzfira et al. 2005). Nevertheless, VirE2 may also function in targeting the T-complex to points of integration through interaction with the host protein VIP1. VIP1 already interacts with VirE2 in the cytoplasm (Djamei et al. 2007; Tzfira et al. 2001) and acts as a mediator between VirE2 and the host nuclear-import machinery (Tzfira et al. 2002). VIP1 has also been implicated in decondensation of the plant chromatin (Avivi et al. 2004) and its interactions with histone proteins (Li et al. 2005a; Loyter et al. 2005) suggest that it may also function in targeting the T-complex to chromatin. More specifically, VIP1 was reported to interact with four core histones of *Xenopus* (Li et al. 2005a) and with the *Arabidopsis* core histone H2A (Li et al. 2005a; Loyter et al. 2005). The latter has been shown to be essential for T-DNA integration in *Arabidopsis* plants (Mysore et al.

2000b). Further investigation of the VIP1-H2A molecular link revealed that an *Arabidopsis* mutant, capable of producing a truncated VIP1 protein, was still capable of supporting nuclear import, but not integration, of the T-DNA (Li et al. 2005a). That truncated VIP1 was unable to interact with H2A, but could still bind to VirE2, in planta (Li et al. 2005a), further supporting the notion that VIP1 acts as a linker between the T-complex and plant chromatin.

It is likely that VirE2 and VirD2 are removed from the T-strand prior to or during the actual integration step. The mechanism that removes VirD2 from the T-strand is still unknown, but it was recently shown that the host-targeted proteolysis machinery may function to strip the T-complex of VirE2 (Tzfira et al. 2004b). More specifically, it was shown that VirF, an F-box-containing bacterial protein which can be transported from the bacterium into the host-cell nucleus (Table 1) (Schrammeijer et al. 2001; Tzfira et al. 2004b; Vergunst et al. 2005), interacts with VIP1 and ASK1. ASK1 is a component of the ASK1-Cdc53-cullin-F-box (SCF) complex which is involved in protein degradation. Using yeast and plant-based functional assays, it was demonstrated that this SCF complex is most likely responsible for stripping the T-strand of VIP1 and VirE2 (Tzfira et al. 2004b). The fact that a mutation in an F-box-like gene in *Arabidopsis* hindered the plant's susceptibility to *Agrobacterium*-mediated transformation (Zhu et al. 2003a) further supports the notion that the host-targeted proteolysis machinery may be involved in the transformation process, perhaps by targeting the T-complex-escorting proteins to degradation.

3.2.2 Functions of Host Proteins in the Integration Process

In addition to the above-mentioned host factors which function through interactions with VirD2 and VirE2, other host proteins (Table 1) have been implicated in the integration process (reviewed in Tzfira et al. 2004a; Ziemienowicz et al. 2008). Host factors are most likely to act in converting the naked (or partially naked) T-strand substrate into a double-stranded form, by providing breaks or nicks in the plant genome into which T-DNA molecules can integrate, and by incorporating T-DNA molecules into the plant genome. Much of what we know today about the functions of host DNA-repair proteins in the integration process are derived from the use of yeast-based transformation systems. These systems allow directing the T-DNA to integration via either homologous recombination (HR) (Bundock et al. 1995; Risseeuw et al. 1996) or nonhomologous end joining (NHEJ) (Bundock and Hooykaas 1996) pathways. More specifically, when T-DNA molecules are homologous to the yeast genome, they integrate via HR (Bundock et al. 1995), while in the absence of such homology, integration occurs via NHEJ (Bundock and Hooykaas 1996). These observations suggest that T-DNA integration is governed predominantly by host factors. Indeed, yeast-mutant analyses revealed that Ku70, Rad50, Mre11, Xrs2, Lig4, and Sir4 are all required for NHEJ-mediated T-DNA integration (van Attikum et al. 2001), while Rad51 and Rad52, but not Rad50, Mre11, Xrs2, Lig4, or Ku70 are required for HR-mediated T-DNA integration (van Attikum and

Hooykaas 2003). Furthermore, since *Ku70* or *Rad52* mutants were deficient in T-DNA integration via NHEJ or HR, respectively, and since a *Ku70/Rad52* double mutant was blocked in T-DNA integration (van Attikum et al. 2001; van Attikum and Hooykaas 2003), Ku70 and Rad52 were suggested to function as the key enzymes determining the route for T-DNA integration in yeast cells.

In contrast to yeast and regardless of the T-DNA sequence, T-DNA integration in plant species occurs predominantly by NHEJ. This phenomenon suggests that NHEJ proteins are the main players in T-DNA integration in plant species. The search for such factors yielded a collection of DNA-repair and maintenance proteins that may function at various steps and by various mechanisms during the integration process. Two *Arabidopsis* mutants, *rad5* and *uvh1*, were perhaps the first plants to be reported as deficient in T-DNA integration due to a possible single-gene mutation (Sonti et al. 1995). Both mutants were hypersensitive to radiation and while *uvh1* could still be stably transformed by *Agrobacterium* (Chateau et al. 2000; Nam et al. 1998; Preuss et al. 1999), its hypersensitivity to the DNA-damaging antibiotic belomycin and the homology between *Arabidopsis RAD5* and the yeast *RAD51* genes suggest a role for host DNA-repair proteins in T-DNA integration in plant cells.

Use of a root-based infection assay, in which plants are screened for their resistance to *Agrobacterium*-mediated transformation (*rat* mutants), led to the identification of several mutants which may be blocked at later stages of the transformation process, including T-DNA integration (Nam et al. 1999; Zhu et al. 2003b). One such mutant, *rat5*, was knocked out in the histone *H2A-1* gene and was indeed blocked in T-DNA integration (Mysore et al. 2000b). Interestingly, *rat5* was still susceptible to *Agrobacterium* transformation via flower infiltration (Mysore et al. 2000a), which suggested that different host factors and mechanisms may function during T-DNA integration in different tissues and/or developmental stages of the cell. Further investigation into the functions of H2A during the integration process revealed that wounding or exposure of root segments to plant-growth regulators increased not only the transformation efficiency but also *H2A-1* expression (Yi et al. 2002). Interestingly, the *rat5* phenotype could be complemented by overexpression of various other *H2A* genes, but only by *H2A-1* when expressed under native promoters (Yi et al. 2006). Thus, while *H2A-1* expression may not necessarily be strictly linked to the cell's S-phase, *H2A-1* expression may serve as a marker for the cell's susceptibility to *Agrobacterium* transformation. The exact molecular function of H2A in the integration process is still largely unknown, but as already described, H2A may function in targeting the T-complex to the plant chromatin through interactions with VIP1 (Li et al. 2005a; Loyter et al. 2005). Interestingly, reduced transcript levels of several histone genes have also been observed in VirE2-interacting protein 2 (VIP2) mutant *Arabidopsis* plants (Anand et al. 2007). Since VIP2 was reported to be essential for T-DNA integration, and since it was also capable of interacting not only with VirE2 but also with VIP1, a possible link between a putative T-DNA-VirE2-VIP1-VIP2 complex and plant chromatin can be suggested (Anand et al. 2007). It should also be noted that downregulation of several other chromatin components also yielded plants with reduced competence toward *Agrobacterium* transformation (Yi et al. 2006), which further stresses the important role of chromatin structure, packaging and remodeling in the integration process (Gelvin and Kim 2007). Another interesting functional link between chromatin remodeling and T-DNA integration was recently reported by Endo et al. (2006), who discovered that mutants in one of the three subunits of the chromatin assembly factor (CAF) show increased T-DNA integration. The authors suggested that in the absence of CAF, T-DNAs are more accessible to the chromosomal DNA and thus more prone to integration. Furthermore, since CAF functions in DNA replication, nucleotide-excision repair and HR (Endo et al. 2006; Kirik et al. 2006), it was also suggested that CAF may actually function in inhibiting T-DNA integration by protecting the chromosomal DNA. It should be noted that the efficiency of T-DNA integration was not affected in *Arabidopsis* mutants which were affected in INO80, another chromatin remodeling complex subunit that is specific to HR (Fritsch et al. 2004), suggesting that not only specific DNA structure, but also specific DNA-repair mechanisms and proteins dictate the integration process.

Various studies have shown that genomic double-strand breaks (DSBs) can serve as hot spots for the integration of dsT-DNA and possibly ssT-DNA molecules (Chilton and Que 2003; Salomon and Puchta 1998; Tzfira et al. 2003). Since both DSBs and T-DNA integration are dominated by NHEJ, understanding the plant NHEJ pathway may provide additional clues to the role of plant factors in the integration process. NHEJ requires the activity of a specific set of proteins, which in yeast include Ku70, Ku80, Rad50, Mre11, Xrs2, Lif1, Nej1, Lig4, and Sir4 (Haber 2000). In mammals, and possibly plants as well, NHEJ repair of DSBs involves the use of KU80, KU70, DNA-PKcs, XRCC4, DNA ligase IV, and MRE11-RAD50-NBS1 (MRN) complex (Bray and West 2005; Weterings and van Gent 2004), suggesting that T-DNA integration into DSBs requires the combined action of many DNA-repair proteins. Indeed, KU80 has recently been shown to be required for T-DNA integration in somatic cells and to physically bind to dsT-DNA molecules in planta (Li et al. 2005b). The role of KU80 in germ-line transformation remains inconclusive since the loss of KU80 expression had only a negligible effect on overall transformation efficiency via the flower-dip method (Friesner and Britt 2003; Gallego et al. 2003).

Additional information about the role of plant ligases in the integration process has been culled from studying VirD2 functions. More specifically, biochemical studies revealed that VirD2 is not capable of ligating T-DNA molecules to acceptor plant DNA in an in-vitro ligation assay, and it was therefore suggested that a plant DNA ligase(s) may be required for ligating T-DNAs to the host genome (Ziemienowicz et al. 2000). Furthermore, since at least in vitro, T-DNA molecules can be ligated to acceptor DNA using various types of DNA ligases (Pansegrau et al. 1993), it may be that the involvement of a particular DNA ligase in the integration process is determined by the type of T-DNA substrate, its integration pathway, and the type of cell in which it integrates. Two different plant DNA ligases, AtLIG1 and AtLIG4, have been studied for their possible role in T-DNA integration. Since the At*LIG1* homozygous mutation is lethal (Babiychuk et al. 1998), this ligase's function could not be analyzed in living plants. However, biochemical studies using an in-vitro integration assay showed that AtLIG1 is capable of interacting with

VirD2 and of facilitating the ligation of T-DNA molecules into model chromosomal DNA (Wu 2002). The role of AtLIG4 in T-DNA integration is still controversial: one study has shown that it is required for germ-line transformation (Friesner and Britt 2003) while another report found it dispensable for T-DNA integration using both root and germ-line transformation (van Attikum et al. 2003). It should be noted, however, that these discrepancies may be attributed to the nature of the flower-dip transformation method.

In addition to KU80, H2A-1, and DNA ligases, other plant DNA-repair and maintenance proteins may also function during the integration process. Plant DNA polymerase(s), for example, are required for the conversion of ssT-DNA to dsT-DNA during or prior to its integration. Others may include additional DNA-repair and/or recombination proteins and further investigation is needed for their identification and to determine their role in the integration process.

3.2.3   The Host DNA Structure Affects T-DNA Integration

T-DNA molecules encounter highly complex genomic DNA structures upon their arrival at the host chromosome. T-DNA molecules, at least in principle, can integrate into every genomic location. This raises the question of whether recipient genomic DNA structures influence the access of T-DNA molecules (and their escorting proteins) to specific points of integration, at the level of the gene and the chromosome. Early reports suggested that T-DNA molecules preferentially integrate into transcriptionally active genomic regions (Tinland 1996; Tinland and Hohn 1995). The fact that integration of promoter-activating trap-like T-DNA molecules often yielded transgenic plants in which a promoterless gene integrated near an active genomic promoter (Herman et al. 1990; Kertbundit et al. 1991; Koncz et al. 1989) further supported the notion that T-DNA molecules are directed to specific genomic sites and perhaps into active genes. Analysis of large-scale T-DNA insertion collections in *Arabidopsis* revealed that the distribution of T-DNA insertions correlated with gene distribution across all five chromosomes, indicating a potential preference for T-DNA insertion into gene sequences (Alonso et al. 2003; Brunaud et al. 2002; Forsbach et al. 2003; Rosso et al. 2003; Schneeberger et al. 2005; Sessions et al. 2002; Szabados et al. 2002). Furthermore, data compiled from several libraries in *Arabidopsis* and rice indicated high T-DNA insertion frequency at the gene's transcription initiation and termination sites, as compared to T-DNA insertions within a gene-coding sequence and intron regions (Chen et al. 2003; Li et al. 2006; Schneeberger et al. 2005; Zhang et al. 2007). While these observations supported the notion that T-DNAs integrate into transcriptionally active sites, correlating T-DNA insertion with transcriptional activity proved to be difficult. On the one hand, Schneeberger et al. (2005) revealed a positive correlation between the frequency of T-DNA insertions in the 5′ upstream regions and gene expression; on the other, Alonso et al. (2003) did not correlate gene-transcription levels with insertion of T-DNA molecules to points of transcription initiation and termination.

Thus, while transcription activity can potentially be correlated with T-DNA insertion into specific sites, it may also be that chromatin configuration, rather than transcription itself, determines T-DNA integration at specific locations. It should also be noted that analyzing the transcript levels in female gametophytes, the target tissue for *Agrobacterium* transformation using the flower-dip method, is rather difficult and can lead to a potentially biased analysis in evaluating the precise role of transcription activity in T-DNA integration.

Analyzing T-DNA insertions at the chromosome level in *Arabidopsis* plants revealed a higher frequency of T-DNA insertions in gene-rich regions, relative to centromeric, paracentromeric, and telomeric sequences which were found to carry less T-DNA insertions (Alonso et al. 2003; Brunaud et al. 2002; Rosso et al. 2003; Sessions et al. 2002; Szabados et al. 2002). A similar distribution pattern was also reported for the rice genome, which has a much greater genomic complexity (An et al. 2003; Chen et al. 2003; Sallaud et al. 2004 Zhang et al. 2007). It thus seems that T-DNA integration in rice and *Arabidopsis*, two species with different genome complexities and different natural responses and susceptibilities to *Agrobacterium* infection, are influenced by similar factors. The fact that T-DNA insertion sites are flanked with AT-rich regions suggests that the host DNA is more flexible, bendable and possibly breakable at points of preintegration, which may promote the access of T-DNA molecules and DNA-repair machineries (Brunaud et al. 2002).In deed, a correlation between predicted DNA bendability at preintegration sites and the frequency of T-DNA integration into such sites has been reported in both *Arabidopsis* and rice plants (Schneeberger et al. 2005; Zhang et al. 2007).

An important question that needs to be asked when analyzing the data obtained from large-scale T-DNA integration studies is whether the genetic material, typically in the form of fully functional transgenic plants, truly represents the population of all possible integration events. Since the regeneration and/or selection of transgenic plants requires the expression of a selectable marker located in the integrating T-DNA molecule, the data are likely to be biased against T-DNA integration into silenced genomic regions which may fail to be selected or regenerate into mature transgenic plants. Indeed, by recovering transgenic plants from two parallel experiments which were performed with or without the application of selectable pressure, Francis and Spiker (2005) revealed that nearly 30% of the plants recovered in the absence of selectable pressure could not be recovered if selection was applied. Furthermore, T-DNA insertions in many transgenic plants have been mapped to genomic regions which were significantly under-represented in selection-based experiments (Francis and Spiker 2005). Similarly, Kim et al. (2007) also analyzed T-DNA integration under nonselective conditions and revealed that about 10% of the T-DNA molecules integrated into centromeric and telomeric sequences of the *Arabidopsis* genome. Taken together, these studies suggest that T-DNA integration can certainly occur in regions which are under-represented in transgenic T-DNA insertion lines, and that T-DNA integration may be more random than has been concluded from large-scale studies of T-DNA insertion collections.

3.2.4 Genomic DSBs and Their Role in the Integration Process

T-DNA integration sometimes results in the formation of complex T-DNA insertion (i.e., the insertion of multiple T-DNA molecules arranged in various orientations relative to each other) and in the insertion of filler DNA between the T-DNA(s) and the host genome (reviewed in Tzfira et al. 2004a; Windels et al. 2008). Among the different models proposed to explain the formation of complex integration patterns (reviewed in Tzfira et al. 2004a; Windels et al. 2008), the possibility that T-DNA molecules recombine to each other prior to their final integration may be the simplest route for the formation of such complex patterns. Since T-DNA molecules cannot recombine head-to-head or tail-to-tail when they are in a ssDNA form, it was suggested that these molecules are first converted to double-stranded intermediates and only then recombine and integrate into the genome (De Buck et al. 1999; De Neve et al. 1997). In this scenario, dsT-DNA molecules are probably recognized by the host DNA-repair machinery as genomic DSBs and recombined to each other via the NHEJ pathway. This notion is supported by the observations that overexpression of KU80, a key protein in the NHEJ DNA-repair pathway, increases T-DNA to T-DNA recombination and T-DNA integration (Li et al. 2005b). It is likely that dsT-DNA molecules also function as intermediates not only for complex, but also for single T-DNA insertions, and that the host NHEJ machinery leads them to integration into genomic DSBs. The possibility that DSBs are involved in T-DNA integration and models for the integration of either ss or dsT-DNA molecules into these breaks were indeed proposed (De Neve et al. 1997; Mayerhofer et al. 1991).

The role of DSBs in T-DNA integration was further investigated in tobacco plants by expressing rare-cutting restriction enzymes capable of inducing DSBs at predetermined genomic sites (Chilton and Que 2003; Salomon and Puchta 1998; Tzfira et al. 2003). DSB induction in transgenic plants resulted in frequent incorporation of new T-DNA molecules into the break sites following a second transformation cycle (Chilton and Que 2003; Salomon and Puchta 1998; Tzfira et al. 2003). Incorporating a recognition site for the rare-cutting restriction enzyme on the incoming T-DNA, and not only on the target plant DNA resulted, in some cases, in the integration of truncated T-DNA molecules which had been digested, in planta, prior to their integration (Chilton and Que 2003; Tzfira et al. 2003). These experiments revealed intriguing aspects of the T-DNA integration process. First, since T-DNA molecules can only be digested by rare-cutters as double-stranded intermediates, the integration of digested T-DNAs provided direct evidence that T-DNA molecules can indeed be converted into dsT-DNA intermediates prior to their integration into the genome. Second, the precise ligation of several digested T-DNA molecules into the genomic DSBs suggests that integration of dsT-DNA molecules may be governed by a simple ligation-like process. Third, because T-DNA molecules are preferentially integrated into rare-cutter-induced genomic DSBs (as determined using nonselective conditions, Tzfira et al. 2003), naturally occurring DSBs might be the driving force for T-DNA integration. The latter is supported by observations that the use of X-ray irradiation, which is known to cause genomic DSBs, can enhance T-DNA integration (Kohler et al. 1989) and that multiple T-DNA molecules, even when delivered from different *Agrobacterium* cells, can integrate into the same genomic locus (De Block and Debrouwer 1991; De Neve et al. 1997; Krizkova and Hrouda 1998).

It is still not known whether *Agrobacterium* can actually induce DSBs at the host genome or whether it relies solely on the natural occurrence of such breaks for the integration of its T-DNA. Nevertheless, that T-DNA molecules can be targeted to genomic DSBs opens the possibility of developing methods for controlled T-DNA integration in plant species, as described later.

4 Models for T-DNA Integration in Plant Cells

Combining our current knowledge on the role of various host and bacterial proteins and the different possible substrates used by these proteins during the integration process into a single T-DNA integration model is perhaps an impossible task. In fact, *Agrobacterium* may use various DNA-repair pathways and different host factors for the integration of either ss or dsT-DNA intermediates in various cell lines and plant species. Indeed, different models have been suggested to explain the integration of single and complex T-DNA integration patterns in plant species (reviewed in Tzfira et al. 2004a; Windels et al. 2008; Ziemienowicz et al. 2008). Lack of space prevents us from discussing all the possible integration models, especially those which describe the integration of complex T-DNA molecules. These models have been discussed in several recent reviews (e.g., Tzfira et al. 2004a; Windels et al. 2008; Ziemienowicz et al. 2008) and we therefore focus here on two principle models which can explain the integration of ss or dsT-DNA molecules into the plant genome.

4.1 Early T-DNA Integration Models

Early models for T-DNA integration were based on analyzing the products of successful integration events, i.e., T-DNA/plant junctions in transgenic plants (Gheysen et al. 1991; Mayerhofer et al. 1991). Data obtained from the sequencing of a very small number of T-DNA inserts revealed that T-DNA molecules lose part of their original sequences. More specifically, T-DNA molecules lost a few nucleotides at their 3′ end, but usually maintained their 5′ end intact (Gheysen et al. 1991; Mayerhofer et al. 1991). Furthermore, small deletions were also observed at the T-DNA's preintegration sites and further analysis revealed a certain homology between the T-DNA's original ends and the preintegration sites. It was thus suggested that homology between the T-strand and the host genome may direct or assist with the integration process. Furthermore, since the homology between the

T-DNA and the plant genome was higher at its 3′ end than at its 5′ end, and since the integration at the T-DNA's 3′ end was less precise than at its 5′ end, the authors suggested that each end of the T-DNA plays a distinct role in the integration process. More specifically, the authors suggested two possible models, designated the "double-strand-break repair" (DSBR) and the "single-strand-gap repair" (SSGR), to describe the integration of ds and ssT-DNA molecules, respectively (Gheysen et al. 1991; Mayerhofer et al. 1991).

The DSBR model assumes that a DSB occurs in the host genome prior to integration. It also suggests that the T-DNA substrate for integration is in a double-stranded form. According to this model, T-DNA integration begins with annealing of the dsT-DNA's ends with the broken host DNA, which may have already lost a few nucleotides due to the activity of host exonucleases. Next, the dsT-DNA overhangs are removed by host exonucleases and/or endonucleases and the trimmed dsT-DNA is then ligated to the host genome. The SSGR model, on the other hand, relies on the presence of gaps in the host genome as the driving force for integration. Such gaps may initiate from naturally occurring nicks which were extended by a 5′->3′ endonuclease. The model also assumes that the T-DNA substrate for integration is a single-stranded molecule. According to this model, integration begins with annealing of the ssT-DNA's ends and the host genome, followed by trimming of the T-DNA's 3′ overhang by host exonucleases and/or endonucleases. Next, the ssT-DNA's 5′ end is ligated to the host genome (originally proposed to be mediated by VirD2) and a second nick in the genome's complementary strand occurs. Integration is completed upon complementation of the T-strand to a double-stranded molecule and ligation of its trimmed 3′ end to the second nick.

It is important to note the fundamental differences in these two models. First, the DSBR integration model assumes that the integrating substrate is a dsT-DNA molecule, while the SSGR model suggests that the integration substrate is in fact an ssT-DNA molecule. Second, the DSBR model requires the occurrence of a genomic DSB, while the SSGR model assumes that a gap or nick in the host genome is sufficient for T-DNA integration. While the integration of dsT-DNA molecules into genomic DSBs was never ruled out, various observations favored the SSGR integration model. These included the fact that T-DNAs are transferred into the host cell as single-stranded molecules and the notion that VirD2 acts as a DNA ligase in living cells. Furthermore, the high transformation efficiency of artificial ssDNA molecules as compared with that of dsDNA molecules (Rodenburg et al. 1989) and the precise integration at the T-DNA's 5′ end as compared with its 3′ end further supported the notion that T-DNA molecules integrate as single-stranded intermediates. These observations led to the establishment of the SSGR, and its derivative model – the microhomology-based T-DNA integration model (see below) – as the dominant model for T-DNA integration. It should be noted, however, that while it was suggested that VirD2 is directly involved in the integration process by ligating the T-DNA into the host genome (Pansegrau et al. 1993), this function was later disputed (Ziemienowicz et al. 2000).

4.2 The Microhomology-Based T-DNA Integration Model

Two major observations led to the establishment of the SSGR as the dominant model for T-DNA integration in plant cells. First, VirD2 is capable of not only digesting, but also re-joining single-stranded substrates in vitro (Pansegrau et al. 1993). This led to the suggestion that VirD2 may function as the T-DNA ligase in plant cells, and by implication, that T-DNAs integrate as single-stranded molecules. Second, specific mutations at the VirD2 putative integrase motif resulted in loss of T-DNA-integration precision, but not of integration efficiency (Tinland et al. 1995). Since the loss of precision resulted in small deletions at the 5′ end of the T-DNA molecules, Tinland et al. (1995) revisited the SSGR integration model to include a specific role for VirD2 during the integration process. According to this model (Fig. 2), only short sequences of the T-DNA molecule actually anneal to preintegration genomic sites. These regions of microhomology were proposed to be the driving force behind the integration process since the model also assumes that the host DNA is only unwound, and not necessarily nicked or broken, at the preintegration sites. The microhomology-based T-DNA integration model also required the function of host endonucleases in producing genomic nicks and in removing the T-strand's 3′ end during the integration process (Fig. 2). It should be noted that while a recent report has shown that VirD2 does not possess DNA-ligase activity (Ziemienowicz et al. 2000), the microhomology-based T-DNA integration model may still be valid. It does, however, require small revisions, in which VirD2 may not act as a DNA ligase, but may still function in the integration process by recruiting a plant ligase(s) and/or other DNA-repair proteins to the integration site (Ziemienowicz et al. 2008).

4.3 Integration of dsT-DNA into Genomic DSBs

The formation of certain complex T-DNA integration patterns, the integration of T-DNA molecules arranged in the same orientation relative to one another and the presence of filler DNA between integrating T-DNA molecules and between T-DNAs and the plant genome (e.g., De Buck et al. 1999; De Neve et al. 1997; Journin et al. 1989; Krizkova and Hrouda 1998) cannot all be simply explained by the microhomology-based T-DNA integration model. Indeed, various modifications to this model, as well as other more unique models have been proposed in several papers in attempts to describe the formation of complex T-DNA integration patterns (reviewed in Tzfira et al. 2004a; Windels et al. 2008; Ziemienowicz et al. 2008).

Sequence analysis of multiple T-DNA molecule insertions at the same genomic locations revealed, in some cases, precise fusion between two right-border ends (De Buck et al. 1999; De Neve et al. 1997). It was thus suggested that since two T-strands cannot recombine at their right borders, these molecules must have been converted to double-stranded intermediates and recombined prior to their integration into the host genome (De Buck et al. 1999; De Neve et al. 1997). More specifically,

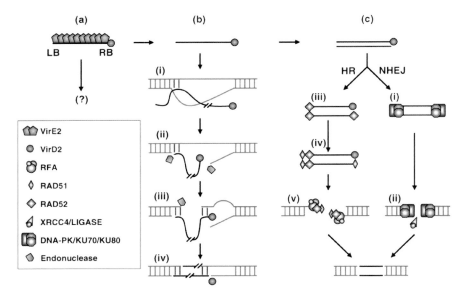

Fig. 2 Possible routes for T-DNA integration in plant cells. (**a**) T-DNA molecules can potentially be directed to integration as mature T-complexes. However, the lack of experimental evidence does not allow modeling this integration route. (**b**) T-strands are the substrate of choice for the microhomology-dependent integration model. According to this model, integration begins with the annealing of the T-strand's 3′ end to the host genome (step i); overhangs of the genomic DNA's bottom strand and the T-strand's 3′ end are trimmed by putative plant endo/exonucleases (step ii). The process continues with annealing of the T-strand's 5′ end to the genomic top strand and nicking of the latter (step iii). The nick is extended into a gap, the T-strand is ligated to the host genome, VirD2 is released and the T-strand is complemented into a double-stranded form (step iv). (**c**) dsT-DNA molecules are the substrate of choice for the DSB-dependent integration model in which integration occurs via NHEJ. According to this model, VirD2 is released from dsT-DNA molecules and bound by the KU70/80/DNA-PK complex (step i), which protects them and leads them to integration into genomic DSBs with the assistance of the XRCC4/LIGASE complex (step ii). dsT-DNA molecules can also potentially be led to integration via HR. In this scenario, dsT-DNA intermediates may be recognized by HR DNA-repair machinery (composed of RAD-like plant proteins, steps iii and iv) and be directed to integration with the assistance of a putative RFA/RAD complex (step v). Illustrations adapted with permission from Tzfira et al. (2004a)

it was suggested that recombination of dsT-DNA molecules to each other as well as their integration into the host genome may be achieved by a host-dependent ligation mechanism (De Buck et al. 1999; De Neve et al. 1997). In such a scenario, VirD2 may still remain attached to the T-DNA molecules, even after their conversion to double-stranded intermediates, but it does not necessarily function as a DNA ligase in planta. Thus, the various models proposed to explain the integration of complex T-DNA structures provide a basis for the idea that T-DNA molecules can be converted to dsT-DNA intermediates prior to their integration. For further information on complex T-DNA integration patterns, the reader is referred to Windels et al. (2008).

The different models for complex T-DNA integration (Windels et al. 2008) do not provide a simple explanation for the incorporation of filler DNA that is sometimes associated with T-DNA integration (e.g., Bakkeren et al. 1989; Gheysen et al. 1991; Journin et al. 1989; Mayerhofer et al. 1991). Filler DNA has also been reported at integration sites of double-stranded plasmid DNA into the plant genome (Gorbunova and Levy 1997). Filler DNA can derive from scrambled plant and/or foreign DNA molecules and its occurrence is often associated with activity of the host NHEJ machinery during the repair of genomic DSBs. Using a functional assay by which DSBs can be induced in plant cells via expression of a rare-cutting restriction enzyme, the repair of these DSBs was associated not only with deletions and insertions of filler DNA molecules, but also with the incorporation of T-DNA molecules (Salomon and Puchta 1998). It was thus suggested that DSBs may play a role in the T-DNA integration and that the random and low-frequency occurrence of such DSBs in the host genome are the limiting factors for T-DNA integration in plant cells (Salomon and Puchta 1998). Further investigation into the structure of T-DNA integration into genomic DSBs revealed that at least in some cases, it was dsT-DNA molecules, and not ssT-DNAs, that integrated into the plant genome (Chilton and Que 2003; Tzfira et al. 2003). It is thus likely that DSBs function as hot spots for the integration of dsT-DNA (and perhaps also ssT-DNA) molecules and that T-DNA integration is governed by the plant NHEJ machinery. The facts that X-ray radiation can enhance transgene integration (Kohler et al. 1989) and that *ku80* mutant *Arabidopsis* plants are resistant to T-DNA integration (Li et al. 2005b) further support the idea that DSBs and the host NHEJ machinery play an important role in T-DNA integration.

A model can be suggested to describe the integration of dsT-DNA substrates via NHEJ into genomic DSBs (Fig. 2). According to this model, the T-strand, which could be degenerated at its unprotected 3′ end, is first converted into a double-stranded intermediate. The dsT-DNA intermediate can now, at least theoretically, be led to integration via HR or NHEJ. Since DSBs are often repaired by NHEJ in somatic cells of many plant species, most T-DNA molecules are likely to be directed to integration by the NHEJ machinery. Thus, components of the NHEJ (e.g., KU80) may function by replacing VirD2, stabilizing the dsT-DNA and leading it to integration into the broken genome. This model can also potentially explain the integration of multiple T-DNA molecules into the same genomic sites, since several dsT-DNA molecules can now be ligated together before or during their integration into the break site (reviewed in Lacroix et al. 2008; Tzfira et al. 2004a). It should be noted that while the integration of dsT-DNA molecules into genomic sites is certainly one mode of T-DNA entry into the plant genome, further research is needed to determine whether it is a major or minor pathway for T-DNA integration in plant cells.

5 Site-Specific T-DNA Integration and GT in Plants

The host DNA-repair machinery dictates the pathway by which T-DNA molecules are recognized and directed to integration. In yeast cells, T-DNA molecules can be directed by either HR or NHEJ DNA-repair pathways, depending on the yeast's genetic makeup and the sequence of the T-DNA molecule (van Attikum et al. 2001;

van Attikum and Hooykaas 2003). In plant species, however, the dominance of NHEJ over HR (Britt and May 2003; Ray and Langer 2002) does not permit the integration of T-DNA molecules via HR and thus hinders our ability to develop efficient GT methods for these species. In fact, only a few reports have described successful GT by T-DNA molecules. This was achieved by developing ingenious vectors and selection schemes for the identification and selection of the rare HR-mediated T-DNA integration events in plants (e.g., Hanin et al. 2001; Terada et al. 2002; Terada et al. 2007). For example, GT in rice has been accomplished using a combination of optimized *Agrobacterium*-mediated genetic transformation and a unique positive and negative selection which eliminated the regeneration of transgenic plants carrying random T-DNA integration events (Terada et al. 2002; Terada et al. 2007). In other examples, T-DNA molecules have been directed for site-specific integration by increasing the rate of HR using site-specific recombinases (Vergunst and Hooykaas 1998; Vergunst et al. 1998). Since T-DNA molecules can be directed for integration into genomic DSBs, rare-cutting restriction enzymes can potentially be used for targeting T-DNA molecules into specific genomic sites. Indeed, both I-*Sce*I and I-*Ceu*I meganucleases have been successfully used to induce site-specific genomic breaks, digest incoming dsT-DNA molecules and target them to predetermined genomic sites (Chilton and Que 2003; Tzfira et al. 2003). It should be noted, however, that just like with site-specific recombinases, the use of meganucleases is limited to targeting new T-DNA molecules into genomic sites which have been previously engineered in the plant genome.

The high efficiency of meganuclease-mediated site-specific insertion of T-DNA molecules in plants (Chilton and Que 2003; Tzfira et al. 2003) suggests that methods which rely on the induction of site-specific DSBs may be useful for GT in plant species. Recent advances in the design and construction of zinc-finger nucleases (ZFNs) as custom-made artificial restriction enzymes and their application for the induction of genomic DSBs in various species (Bibikova et al. 2003; Moehle et al. 2007; Porteus 2006), including plants (Durai et al. 2005; Lloyd et al. 2005; Wright et al. 2005), may offer a new way of harnessing one of T-DNA's integration routes for GT in plants. Since ZFNs can be tailored to target various sequences, they offer a unique opportunity for targeting native genomic sequences in the plant genome. Indeed, the application of ZFNs in plant species has been demonstrated in *Arabidopsis* where the induced expression of a ZFN under the control of a heat-shock promoter led to the induction of site-specific DSBs to site-specific mutagenesis of an artificial target site (Lloyd et al. 2005). While the authors did not report on specific attempts to integrate new T-DNA molecules into these break sites, Wright et al. (2005) showed that T-DNA molecules can indeed be targeted to ZFN-mediated DSBs. It should be noted, however, that in the latter case, T-DNA integration occurred via HR and not NHEJ. Thus, while NHEJ may be the dominant pathway for T-DNA integration, it may be possible to redirect T-DNA molecules to HR via the induction of genomic DSBs. This notion was further supported by observations that induction of genomic DSBs in maize cells by expression of I-*Sce*I resulted in both HR- and NHEJ-mediated insertion of both T-DNA and plasmid DNA into the break sites (D'Halluin et al. 2008).

The route by which T-DNA molecules integrate into the host genome can potentially be influenced not only by the induction of DSBs but also by manipulating the host's genetic markup. In a unique approach, Shaked et al. (2005) enhanced HR-mediated T-DNA integration in *Arabidopsis* plants by overexpressing the gene RAD54, a member of the SWI2/SNF2 chromatin remodeling gene family from yeast. In line with observations that chromatin structure and chromatin-related proteins may affect T-DNA integration in yeast and plant cells, these findings further support the notion that chromatin remodeling dictates the routes for T-DNA integration in plant cells.

6   Conclusions and Future Perspectives

In the past several years, we have come to appreciate how complicated the T-DNA-integration process in plant species really is. We have also revealed this process's heavy reliance on the host DNA's genome structure and host DNA-repair proteins. Early studies and sequencing of large T-DNA insertion collections led to the establishment of various T-DNA integration models. These integration models have been subjected to several revisions in order to accommodate the advances in our understanding of the biochemical role of both bacterial and host proteins in the integration process and these advances would not have been possible without the deployment of a collective effort from various laboratories across the globe that use a wide range of experimental approaches and model systems. These include the development of functional T-DNA integration assays, the use of protein–protein interaction analyses, the identification and characterization of *Arabidopsis* mutants, harnessing the power of yeast genetics and extending the research from *Arabidopsis* to tobacco, rice, and other plant species. Exploring the T-DNA integration process is important not only for a basic understanding of the transformation process but also for the development of new tools and methods that will allow us to control this process in plant species. Indeed, revealing the role of genomic DSBs and the functions of specific chromatin-remodeling proteins in the integration process has already helped in designing basic strategies for GT in plant cells. It is thus rather clear that the identification of additional factors involved in the integration process, revealing the pathways that T-DNA molecules take toward their integration into the host genome and understanding the basic cellular mechanisms by which these molecules integrate into the genome are all needed to further enhance our understanding of the scientifically and agronomically important genetic-transformation process.

Acknowledgments The work in our laboratory is supported by the University of Michigan start-up funds and by grants from the Biotechnology Research and Development Cooperation (BRDC) and the Consortium for Plant Biotechnology Research, Inc. (CPBR), and in collaboration with, and with financial support from DOW Agrosciences.

References

Abu-Arish A, Frenkiel-Krispin D, Fricke T, Tzfira T, Citovsky V, Grayer Wolf S, Elbaum M (2004) Three-dimensional reconstruction of *Agrobacterium* VirE2 protein with single-stranded DNA. J Biol Chem 279:25359–25363

Alonso JM, Stepanova AN, Leisse TJ, Kim CJ, Chen H, Shinn P, Stevenson DK, Zimmerman J, Barajas P, Cheuk R, Gadrinab C, Heller C, Jeske A, Koesema E, Meyers CC, Parker H, Prednis L, Ansari Y, Choy N, Deen H, Geralt M, Hazari N, Hom E, Karnes M, Mulholland C, Ndubaku R, Schmidt I, Guzman P, Aguilar-Henonin L, Schmid M, Weigel D, Carter DE, Marchand T, Risseeuw E, Brogden D, Zeko A, Crosby WL, Berry CC, Ecker JR (2003) Genome-wide insertional mutagenesis of *Arabidopsis thaliana*. Science 301:653–657

An S, Park S, Jeong DH, Lee DY, Kang HG, Yu JH, Hur J, Kim SR, Kim YH, Lee M, Han S, Kim SJ, Yang J, Kim E, Wi SJ, Chung HS, Hong JP, Choe V, Lee HK, Choi JH, Nam J, Park PB, Park KY, Kim WT, Choe S, Lee CB, An G (2003) Generation and analysis of end sequence database for T-DNA tagging lines in rice. Plant Physiol 133:2040–2047

Anand A, Krichevsky A, Schornack S, Lahaye T, Tzfira T, Tang Y, Citovsky V, Mysore KS (2007) *Arabidopsis* VIRE2 INTERACTING PROTEIN2 is required for *Agrobacterium* T-DNA integration in plants. Plant Cell 19:1695–1708

Atmakuri K, Christie PJ (2008) Translocation of oncogenic T-DNA and effector proteins to plant cells. In: Tzfira T, Citovsky V (eds) *Agrobacterium*. Springer, New York, pp 315–364

Avivi Y, Morad V, Ben-Meir H, Zhao J, Kashkush K, Tzfira T, Citovsky V, Grafi G (2004) Reorganization of specific chromosomal domains and activation of silent genes in plant cells acquiring pluripotentiality. Dev Dyn 230:12–22

Babiychuk E, Cottrill PB, Storozhenko S, Fuangthong M, Chen Y, O'Farrell MK, Van Montagu M, Inze D, Kushnir S (1998) Higher plants possess two structurally different poly(ADP-ribose) polymerases. Plant J 15:635–645

Bakkeren G, Koukolikova-Nicola Z, Grimsley N, Hohn B (1989) Recovery of *Agrobacterium tumefaciens* T-DNA molecules from whole plants early after transfer. Cell 57:847–857

Bako L, Umeda M, Tiburcio AF, Schell J, Koncz C (2003) The VirD2 pilot protein of *Agrobacterium*-transferred DNA interacts with the TATA box-binding protein and a nuclear protein kinase in plants. Proc Natl Acad Sci USA 100:10108–10113

Ballas N, Citovsky V (1997) Nuclear localization signal binding protein from *Arabidopsis* mediates nuclear import of *Agrobacterium* VirD2 protein. Proc Natl Acad Sci USA 94:10723–10728

Banta LM, Montenegro M (2008) *Agrobacterium* and plant biotechnology. In: Tzfira T, Citovsky V (eds) *Agrobacterium*. Springer, New York, pp 73–147

Bibikova M, Beumer K, Trautman JK, Carroll D (2003) Enhancing gene targeting with designed zinc finger nucleases. Science 300:764

Bray CM, West CE (2005) DNA repair mechanisms in plants: crucial sensors and effectors for the maintenance of genome integrity. New Phytol 168:511–528

Britt AB, May GD (2003) Re-engineering plant gene targeting. *Trends Plant Sci* 8:90–95

Brunaud V, Balzergue S, Dubreucq B, Aubourg S, Samson F, Chauvin S, Bechtold N, Cruaud C, DeRose R, Pelletier G, Lepiniec L, Caboche M, Lecharny A (2002) T-DNA integration into the *Arabidopsis* genome depends on sequences of pre-insertion sites. EMBO Rep 3:1152–1157

Bundock P, Hooykaas PJJ (1996) Integration of *Agrobacterium tumefaciens* T-DNA in the *Saccharomyces cerevisiae* genome by illegitimate recombination. *Proc Natl Acad Sci USA* 93:15272–15275

Bundock P, den Dulk-Ras A, Beijersbergen A, Hooykaas PJJ (1995) Trans-kingdom T-DNA transfer from *Agrobacterium tumefaciens* to *Saccharomyces cerevisiae*. EMBO J 14:3206–3214

Chateau S, Sangwan RS, Sangwan-Norreel BS (2000) Competence of *Arabidopsis thaliana* genotypes and mutants for *Agrobacterium tumefaciens*-mediated gene transfer: role of phytohormones. J Exp Bot 51:1961–1968

Chen S, Jin W, Wang M, Zhang F, Zhou J, Jia Q, Wu Y, Liu F, Wu P (2003) Distribution and characterization of over 1,000 T-DNA tags in rice genome. Plant J 36:105–113

Chilton M-DM, Que Q (2003) Targeted integration of T-DNA into the tobacco genome at double-strand breaks: new insights on the mechanism of T-DNA integration. Plant Physiol 133:956–965

Christie PJ, Atmakuri K, Krishnamoorthy V, Jakubowski S, Cascales E (2005) Biogenesis, architecture, and function of bacterial type IV secretion systems. *Annu Rev Microbiol* 59:451–485

Citovsky V, Wong ML, Zambryski PC (1989) Cooperative interaction of *Agrobacterium* VirE2 protein with single stranded DNA: implications for the T-DNA transfer process. Proc Natl Acad Sci USA 86:1193–1197

Citovsky V, Zupan J, Warnick D, Zambryski PC (1992) Nuclear localization of *Agrobacterium* VirE2 protein in plant cells. Science 256:1802–1805

Citovsky V, Guralnick B, Simon MN, Wall JS (1997) The molecular structure of *Agrobacterium* VirE2-single stranded DNA complexes involved in nuclear import. *J Mol Biol* 271:718–727

Citovsky V, Kozlovsky SV, Lacroix B, Zaltsman A, Dafny-Yelin M, Vyas S, Tovkach A, Tzfira T (2007) Biological systems of the host cell involved in *Agrobacterium* infection. Cell Microbiol 9:9–20

Dafny-Yelin M, Levy A, Tzfira T (2008) The ongoing saga of *Agrobacterium*-host interactions. Trends Plant Sci 13:102–105

De Block M, Debrouwer D (1991) Two T-DNA's co-transformed into *Brassica napus* by a double *Agrobacterium tumefaciens* infection are mainly integrated at the same locus. Theor Appl Genet 82:257–263

De Buck S, Jacobs A, Van Montagu M, Depicker A (1999) The DNA sequences of T-DNA junctions suggest that complex T-DNA loci are formed by a recombination process resembling T-DNA integration. Plant J 20:295–304

de Cleene M, de Ley J (1976) The host range of crown gall. *Bot Rev* 42:389–466

de Groot MJ, Bundock P, Hooykaas PJJ, Beijersbergen AG (1998) *Agrobacterium tumefaciens*-mediated transformation of filamentous fungi. Nat Biotechnol 16:839–842 [published erratum appears in Nat Biotechnol 16:1074 (1998)]

De Neve M, De Buck S, Jacobs A, Van Montagu M, Depicker A (1997) T-DNA integration patterns in co-transformed plant cells suggest that T-DNA repeats originate from co-integration of separate T-DNAs. Plant J 11:15–29

Deng W, Chen L, Wood DW, Metcalfe T, Liang X, Gordon MP, Comai L, Nester EW (1998) *Agrobacterium* VirD2 protein interacts with plant host cyclophilins. Proc Natl Acad Sci USA 95:7040–7045

D'Halluin K, Vanderstraeten C, Stals E, Cornelissen M, Ruiter R (2008) Homologous recombination: a basis for targeted genome optimization in crop species such as maize. Plant Biotechnol J 6:93–102

Djamei A, Pitzschke A, Nakagami H, Rajh I, Hirt H (2007) Trojan horse strategy in *Agrobacterium* transformation: abusing MAPK defense signaling. *Science* 318:453–456

Duckely M, Hohn B (2003) The VirE2 protein of *Agrobacterium tumefaciens*: the Yin and Yang of T-DNA transfer. FEMS Microbiol Lett 223:1–6

Durai S, Mani M, Kandavelou K, Wu J, Porteus MH, Chandrasegaran S (2005) Zinc finger nucleases: custom-designed molecular scissors for genome engineering of plant and mammalian cells. Nucleic Acids Res 33:5978–5990

Endo M, Ishikawa Y, Osakabe K, Nakayama S, Kaya H, Araki T, Shibahara K, Abe K, Ichikawa H, Valentine L, Hohn B, Toki S (2006) Increased frequency of homologous recombination and T-DNA integration in *Arabidopsis* CAF-1 mutants. EMBO J 25:5579–5590

Forsbach A, Schubert D, Lechtenberg B, Gils M, Schmidt R (2003) A comprehensive characterization of single-copy T-DNA insertions in the *Arabidopsis thaliana* genome. Plant Mol Biol 52:161–176

Francis KE, Spiker S (2005) Identification of *Arabidopsis thaliana* transformants without selection reveals a high occurrence of silenced T-DNA integrations. Plant J 41:464–477

Friesner J, Britt AB (2003) *Ku80*- and *DNA ligase IV*-deficient plants are sensitive to ionizing radiation and defective in T-DNA integration. Plant J 34:427–440

Fritsch O, Benvenuto G, Bowler C, Molinier J, Hohn B (2004) The INO80 protein controls homologous recombination in *Arabidopsis thaliana*. Mol Cell 16:479–485

Gallego ME, Bleuyard JY, Daoudal-Cotterell S, Jallut N, White CI (2003) Ku80 plays a role in non-homologous recombination but is not required for T-DNA integration in *Arabidopsis*. Plant J 35:557–565

Gelvin SB (1998) The introduction and expression of transgenes in plants. Curr Opin Biotechnol 9:227–232

Gelvin SB (2003) *Agrobacterium*-mediated plant transformation: the biology behind the "gene-jockeying" tool. Microbiol Mol Biol Rev 67:16–37

Gelvin SB, Kim SI (2007) Effect of chromatin upon *Agrobacterium* T-DNA integration and transgene expression. Biochim Biophys Acta 1769:410–421

Gheysen G, Villarroel R, Van Montagu M (1991) Illegitimate recombination in plants: a model for T-DNA integration. Genes Dev 5:287–297

Godio RP, Fouces R, Gudina EJ, Martin JF (2004) *Agrobacterium tumefaciens*-mediated transformation of the antitumor clavaric acid-producing basidiomycete *Hypholoma sublateritium*. Curr Genet 46:287–294

Gorbunova V, Levy AA (1997) Non-homologous DNA end joining in plant cells is associated with deletions and filler DNA insertions. Nucleic Acids Res 25:4650–4657

Haber JE (2000) Lucky breaks: analysis of recombination in *Saccharomyces*. Mutat Res 451:53–69

Hanin M, Volrath S, Bogucki A, Briker M, Ward E, Paszkowski J (2001) Gene targeting in *Arabidopsis*. Plant J 28:671–677

Hellens R, Mullineaux P, Klee H (2000) Technical focus: a guide to *Agrobacterium* binary Ti vectors. *Trends Plant Sci* 5:446–451

Herman L, Jacobs A, Van Montagu M, Depicker A (1990) Plant chromosome/marker gene fusion assay for study of normal and truncated T-DNA integration events. Mol Gen Genet 224:248–256

Howard EA, Citovsky V (1990) The emerging structure of the *Agrobacterium* T-DNA transfer complex. BioEssays 12:103–108

Howard EA, Zupan JR, Citovsky V, Zambryski PC (1992) The VirD2 protein of *A. tumefaciens* contains a C-terminal bipartite nuclear localization signal: implications for nuclear uptake of DNA in plant cells. Cell 68:109–118

Journin L, Bouchez D, Drong RF, Tepfer D, Slightom JL (1989) Analysis of TR-DNA/plant junctions in the genome of *Convolvulus arvensis* clone transformed by *Agrobacterium rhizogenes* strain A4. Plant Mol Biol 12:72–85

Kertbundit S, De Greve H, Deboeck F, Van Montagu M, Hernalsteens JP (1991) In vivo random beta-glucuronidase gene fusions in *Arabidopsis thaliana*. Proc Natl Acad Sci USA 88:5212–5216

Kim SI, Veena, Gelvin SB (2007) Genome-wide analysis of *Agrobacterium* T-DNA integration sites in the *Arabidopsis* genome generated under non-selective conditions. Plant J 51:779–791

Kirik A, Pecinka A, Wendeler E, Reiss B (2006) The chromatin assembly factor subunit FASCIATA1 is involved in homologous recombination in plants. Plant Cell 18:2431–2442

Kohler F, Cardon G, Pohlman M, Gill R, Schieder O (1989) Enhancement of transformation rates in higher plants by low-dose irradiation: are DNA repair systems involved in the incorporation of exogenous DNA into the plant genome? Plant Mol Biol 12:189–199

Koncz C, Martini N, Mayerhofer R, Koncz-Kalman Z, Korber H, Redei GP, Schell J (1989) High-frequency T-DNA-mediated gene tagging in plants. Proc Natl Acad Sci USA 86:8467–8471

Krizkova L, Hrouda M (1998) Direct repeats of T-DNA integrated in tobacco chromosome: characterization of junction regions. Plant J 16:673–680

Kunik T, Tzfira T, Kapulnik Y, Gafni Y, Dingwall C, Citovsky V (2001) Genetic transformation of HeLa cells by *Agrobacterium*. Proc Natl Acad Sci USA 98:1871–1876

Lacroix B, Vaidya M, Tzfira T, Citovsky V (2005) The VirE3 protein of *Agrobacterium* mimics a host cell function required for plant genetic transformation. *EMBO J* 24:428–437

Lacroix B, Li J, Tzfira T, Citovsky V (2006a) Will you let me use your nucleus? How *Agrobacterium* gets its T-DNA expressed in the host plant cell. Can J Physiol Pharmacol 84:333–345

Lacroix B, Tzfira T, Vainstein A, Citovsky V (2006b) A case of promiscuity: *Agrobacterium*'s endless hunt for new partners. Trends Genet 22:29–37

Lacroix B, Elbaum M, Citovsky V, Tzfira T (2008) Intracellular transport of *Agrobacterium* T-DNA. In: Tzfira T, Citovsky V (eds) Agrobacterium. Springer, New York, pp 365–394

Li J, Krichevsky A, Vaidya M, Tzfira T, Citovsky V (2005a) Uncoupling of the functions of the *Arabidopsis* VIP1 protein in transient and stable plant genetic transformation by *Agrobacterium*. Proc Natl Acad Sci USA 102:5733–5738

Li J, Vaidya M, White C, Vainstein A, Citovsky V, Tzfira T (2005b) Involvement of KU80 in T-DNA integration in plant cells. Proc Natl Acad Sci USA 102:19231–19236

Li Y, Rosso MG, Ulker B, Weisshaar B (2006) Analysis of T-DNA insertion site distribution patterns in *Arabidopsis thaliana* reveals special features of genes without insertions. Genomics 87:645–652

Lloyd A, Plaisier CL, Carroll D, Drews GN (2005) Targeted mutagenesis using zinc-finger nucleases in *Arabidopsis*. Proc Natl Acad Sci USA 102:2232–2237

Loyter A, Rosenbluh J, Zakai N, Li J, Kozlovsky SV, Tzfira T, Citovsky V (2005) The plant VirE2 interacting protein 1. A molecular link between the *Agrobacterium* T-complex and the host cell chromatin? Plant Physiol 138:1318–1321

Mayerhofer R, Koncz-Kalman Z, Nawrath C, Bakkeren G, Crameri A, Angelis K, Redei GP, Schell J, Hohn B, Koncz C (1991) T-DNA integration: a mode of illegitimate recombination in plants. EMBO J 10:697–704

McCullen CA, Binns AN (2006) *Agrobacterium tumefaciens* plant cell interactions and activities required for interkingdom macromolecular transfer. Annu Rev Cell Dev Biol 22:101–127

Michielse CB, Arentshorst M, Ram AF, van den Hondel CA (2005) *Agrobacterium*-mediated transformation leads to improved gene replacement efficiency in *Aspergillus awamori*. Fungal Genet Biol 42:9–19

Moehle EA, Rock JM, Lee YL, Jouvenot Y, Dekelver RC, Gregory PD, Urnov FD, Holmes MC (2007) Targeted gene addition into a specified location in the human genome using designed zinc finger nucleases. Proc Natl Acad Sci USA 104:3055–3060

Montague JW, Hughes FM Jr, Cidlowski JA (1997) Native recombinant cyclophilins A, B, and C degrade DNA independently of peptidylprolyl cis-trans-isomerase activity. Potential roles of cyclophilins in apoptosis. J Biol Chem 272:6677–6684

Mysore KS, Bassuner B, Deng X-B, Darbinian NS, Motchoulski A, Ream LW, Gelvin SB (1998) Role of the *Agrobacterium tumefaciens* VirD2 protein in T-DNA transfer and integration. Mol Plant-Microbe Interact 11:668–683

Mysore KS, Kumar CT, Gelvin SB (2000a) *Arabidopsis* ecotypes and mutants that are recalcitrant to *Agrobacterium* root transformation are susceptible to germ-line transformation. Plant J 21:9–16

Mysore KS, Nam J, Gelvin SB (2000b) An *Arabidopsis* histone H2A mutant is deficient in *Agrobacterium* T-DNA integration. Proc Natl Acad Sci USA 97:948–953

Nam J, Matthysse AG, Gelvin SB (1997) Differences in susceptibility of *Arabidopsis* ecotypes to crown gall disease may result from a deficiency in T-DNA integration. Plant Cell 9:317–333

Nam J, Mysore KS, Gelvin SB (1998) Agrobacterium tumefaciens transformation of the radiation hypersensitive Arabidopsis thaliana mutants uvh1 and rad5. Mol Plant-Microbe Interact 11:1136–1141

Nam J, Mysore KS, Zheng C, Knue MK, Matthysse AG, Gelvin SB (1999) Identification of T-DNA tagged *Arabidopsis* mutants that are resistant to transformation by *Agrobacterium*. Mol Gen Genet 261:429–438

Otten L, Burr TJ, Szegedi E (2008) *Agrobacterium*: a disease-causing bacterium. In: Tzfira T, Citovsky V (eds) Agrobacterium. Springer, New York, pp 1–46

Pansegrau W, Schoumacher F, Hohn B, Lanka E (1993) Site-specific cleavage and joining of single-stranded DNA by VirD2 protein of *Agrobacterium tumefaciens* Ti plasmids: analogy to bacterial conjugation. Proc Natl Acad Sci USA 90:11538–11542

Piers KL, Heath JD, Liang X, Stephens KM, Nester EW (1996) *Agrobacterium tumefaciens*-mediated transformation of yeast. Proc Natl Acad Sci USA 93:1613–1618

Porteus MH (2006) Mammalian gene targeting with designed zinc finger nucleases. Mol Ther 13:438–446

Preuss SB, Jiang CZ, Baik HK, Kado CI, Britt AB (1999) Radiation-sensitive *Arabidopsis* mutants are proficient for T-DNA transformation. *Mol Gen Genet* 261:623–626

Ray A, Langer M (2002) Homologous recombination: ends as the means. Trends Plant Sci 7:435–440

Risseeuw E, Franke-van Dijk ME, Hooykaas PJ (1996) Integration of an insertion-type transferred DNA vector from *Agrobacterium tumefaciens* into the *Saccharomyces cerevisiae* genome by gap repair. Mol Cell Biol 16:5924–5932

Rodenburg KW, de Groot MJ, Schilperoort RA, Hooykaas PJ (1989) Single-stranded DNA used as an efficient new vehicle for transformation of plant protoplasts. Plant Mol Biol 13:711–719

Rossi L, Hohn B, Tinland B (1996) Integration of complete transferred DNA units is dependent on the activity of virulence E2 protein of *Agrobacterium tumefaciens*. Proc Natl Acad Sci USA 93:126–130

Rosso MG, Li Y, Strizhov N, Reiss B, Dekker K, Weisshaar B (2003) An *Arabidopsis thaliana* T-DNA mutagenized population (GABI-Kat) for flanking sequence tag-based reverse genetics. Plant Mol Biol 53:247–259

Sallaud C, Gay C, Larmande P, Bes M, Piffanelli P, Piegu B, Droc G, Regad F, Bourgeois E, Meynard D, Perin C, Sabau X, Ghesquiere A, Glaszmann JC, Delseny M, Guiderdoni E (2004) High throughput T-DNA insertion mutagenesis in rice: a first step towards in silico reverse genetics. Plant J 39:450–464

Salomon S, Puchta H (1998) Capture of genomic and T-DNA sequences during double-strand break repair in somatic plant cells. EMBO J 17:6086–6095

Schneeberger RG, Zhang K, Tatarinova T, Troukhan M, Kwok SF, Drais J, Klinger K, Orejudos F, Macy K, Bhakta A, Burns J, Subramanian G, Donson J, Flavell R, Feldmann KA (2005) *Agrobacterium* T-DNA integration in *Arabidopsis* is correlated with DNA sequence compositions that occur frequently in gene promoter regions. Funct Integr Genomics 5:240–253

Schrammeijer B, Risseeuw E, Pansegrau W, Regensburg-Tu nk TJG, Crosby WL, Hooykaas PJJ (2001) Interaction of the virulence protein VirF of *Agrobacterium tumefaciens* with plant homologs of the yeast Skp1 protein. Curr Biol 11:258–262

Sessions A, Burke E, Presting G, Aux G, McElver J, Patton D, Dietrich B, Ho P, Bacwaden J, Ko C, Clarke JD, Cotton D, Bullis D, Snell J, Miguel T, Hutchison D, Kimmerly B, Mitzel T, Katagiri F, Glazebrook J, Law M, Goff SA (2002) A high-throughput *Arabidopsis* reverse genetics system. Plant Cell 14:2985–2994

Shaked H, Melamed-Bessudo C, Levy AA (2005) High frequency gene targeting in *Arabidopsis* plants expressing the yeast *RAD54* gene. Proc Natl Acad Sci USA 102:12265–12269

Soltani J, van Heusden GPH, Hooykaas PJ (2008) *Agrobacterium*-mediated transformation of non-plant organisms. In: Tzfira T, Citovsky V (eds) Agrobacterium. Springer, New York, pp 650–674

Sonti RV, Chiurazzi M, Wong D, Davies CS, Harlow GR, Mount DW, Signer ER (1995) *Arabidopsis* mutants deficient in T-DNA integration. Proc Natl Acad Sci USA 92:11786–11790

Szabados L, Kovacs I, Oberschall A, Abraham E, Kerekes I, Zsigmond L, Nagy R, Alvarado M, Krasovskaja I, Gal M, Berente A, Redei GP, Haim AB, Koncz C (2002) Distribution of 1,000 sequenced T-DNA tags in the *Arabidopsis* genome. Plant J 32:233–242

Terada R, Urawa H, Inagaki Y, Tsugane K, Iida S (2002) Efficient gene targeting by homologous recombination in rice. Nat Biotechnol 20:1030–1034

Terada R, Johzuka-Hisatomi Y, Saitoh M, Asao H, Iida S (2007) Gene targeting by homologous recombination as a biotechnological tool for rice functional genomics. Plant Physiol 144:846–856

Tinland B (1996) The integration of T-DNA into plant genomes. Trends Plant Sci 1:178–184

Tinland B, Hohn B (1995) Recombination between prokaryotic and eukaryotic DNA: integration of *Agrobacterium tumefaciens* T-DNA into the plant genome. Genet Eng 17:209–229

Tinland B, Koukolikova-Nicola Z, Hall MN, Hohn B (1992) The T-DNA-linked VirD2 protein contains two distinct nuclear localization signals. Proc Natl Acad Sci USA 89:7442–7446

Tinland B, Schoumacher F, Gloeckler V, Bravo-Angel AM, Hohn B (1995) The *Agrobacterium tumefaciens* virulence D2 protein is responsible for precise integration of T-DNA into the plant genome. EMBO J 14:3585–3595

Tzfira T, Citovsky V (2002) Partners-in-infection: host proteins involved in the transformation of plant cells by *Agrobacterium*. Trends Cell Biol 12:121–129

Tzfira T, Citovsky V (2006) *Agrobacterium*-mediated genetic transformation of plants: biology and biotechnology. Curr Opin Biotechnol 17:147–154

Tzfira T, Vaidya M, Citovsky V (2001) VIP1, an *Arabidopsis* protein that interacts with *Agrobacterium* VirE2, is involved in VirE2 nuclear import and *Agrobacterium* infectivity. EMBO J 20: 3596–3607

Tzfira T, Vaidya M, Citovsky V (2002) Increasing plant susceptibility to *Agrobacterium* infection by overexpression of the *Arabidopsis* nuclear protein VIP1. Proc Natl Acad Sci USA 99:10435–10440

Tzfira T, Frankmen L, Vaidya M, Citovsky V (2003) Site-specific integration of *Agrobacterium tumefaciens* T-DNA via double-stranded intermediates. Plant Physiol 133:1011–1023

Tzfira T, Li J, Lacroix B, Citovsky V (2004a) *Agrobacterium* T-DNA integration: molecules and models. Trends Genet 20:375–383

Tzfira T, Vaidya M, Citovsky V (2004b) Involvement of targeted proteolysis in plant genetic transformation by *Agrobacterium*. Nature 431:87–92

Tzfira T, Lacroix B, Citovsky V (2005) Nuclear Import of *Agrobacterium* T-DNA. In: Tzfira T, Citovsky V (eds) Nuclear Import and Export in Plants and Animals. Landes Bioscience/ Kluwer/Plenum, New York, pp 83–99

van Attikum H, Hooykaas PJJ (2003) Genetic requirements for the targeted integration of *Agrobacterium* T-DNA in *Saccharomyces cerevisiae*. Nucleic Acids Res 31:826–832

van Attikum H, Bundock P, Hooykaas PJJ (2001) Non-homologous end-joining proteins are required for *Agrobacterium* T-DNA integration. EMBO J 20:6550–6558

van Attikum H, Bundock P, Overmeer RM, Lee LY, Gelvin SB, Hooykaas PJ (2003) The *Arabidopsis AtLIG4* gene is required for the repair of DNA damage, but not for the integration of *Agrobacterium* T-DNA. Nucleic Acids Res 31:4247–4255

van Attikum H, Fritsch O, Hohn B, Gasser SM (2004) Recruitment of the INO80 complex by H2A phosphorylation links ATP-dependent chromatin remodeling with DNA double-strand break repair. Cell 119:777–788

Vergunst AC, Hooykaas PJ (1998) Cre/lox-mediated site-specific integration of *Agrobacterium* T-DNA in *Arabidopsis thaliana* by transient expression of cre. Plant Mol Biol 38:393–406

Vergunst AC, Jansen LE, Hooykaas PJ (1998) Site-specific integration of *Agrobacterium* T-DNA in *Arabidopsis thaliana* mediated by Cre recombinase. Nucleic Acids Res 26:2729–2734

Vergunst AC, van Lier MC, den Dulk-Ras A, Grosse Stuve TA, Ouwehand A, Hooykaas PJ (2005) Positive charge is an important feature of the C-terminal transport signal of the VirB/ D4-translocated proteins of *Agrobacterium*. Proc Natl Acad Sci USA 102:832–837

Volokhina I, Chumakov M (2007) Study of the VirE2-ssT-DNA complex formation by scanning probe microscopy and gel electrophoresis – T-complex visualization. Microsc Microanal 13:51–54

Ward DV, Zambryski PC (2001) The six functions of *Agrobacterium* VirE2. Proc Natl Acad Sci USA 98:385–386

Ward D, Zupan J, Zambryski PC (2002) *Agrobacterium* VirE2 gets the VIP1 treatment in plant nuclear import. Trends Plant Sci 7:1–3

Weterings E, van Gent DC (2004) The mechanism of non-homologous end-joining: a synopsis of synapsis. DNA Repair (Amst) 3:1425–1435

Windels P, De Buck S, Depicker A (2008) *Agrobacterium tumefaciens*-mediated transformation: patterns of T-DNA integration into the host genome. In: Tzfira T, Citovsky V (eds) *Agrobacterium*. Springer, New York, pp 441–481

Wright DA, Townsend JA, Winfrey RJ Jr, Irwin PA, Rajagopal J, Lonosky PM, Hall BD, Jondle MD, Voytas DF (2005) High-frequency homologous recombination in plants mediated by zinc-finger nucleases. Plant J 44:693–705

Wu Y-Q (2002) Protein-protein interaction between VirD2 and DNA ligase: an essential step of Agrobacterium tumefaciens T-DNA integration. PhD Thesis, University of Basel

Yi H, Mysore KS, Gelvin SB (2002) Expression of the *Arabidopsis* histone *H2A-1* gene correlates with susceptibility to *Agrobacterium* transformation. Plant J 32:285–298

Yi H, Sardesai N, Fujinuma T, Chan CW, Veena, Gelvin SB (2006) Constitutive expression exposes functional redundancy between the *Arabidopsis* histone *H2A* gene *HTA1* and other *H2A* gene family members. Plant Cell 18:1575–1589

Zambryski PC (1992) Chronicles from the *Agrobacterium*-plant cell DNA transfer story. Annu Rev Plant Physiol Plant Mol Biol 43:465–490

Zhang J, Guo D, Chang Y, You C, Li X, Dai X, Weng Q, Chen G, Liu H, Han B, Zhang Q, Wu C (2007) Non-random distribution of T-DNA insertions at various levels of the genome hierarchy as revealed by analyzing 13 804 T-DNA flanking sequences from an enhancer-trap mutant library. Plant J 49:947–959

Zhu Y, Nam J, Carpita NC, Matthysse AG, Gelvin SB (2003a) *Agrobacterium*-mediated root transformation is inhibited by mutation of an *Arabidopsis* cellulose synthase-like gene. Plant Physiol 133:1000–1010

Zhu Y, Nam J, Humara JM, Mysore KS, Lee LY, Cao H, Valentine L, Li J, Kaiser AD, Kopecky AL, Hwang HH, Bhattacharjee S, Rao PK, Tzfira T, Rajagopal J, Yi H, Veena, Yadav BS, Crane YM, Lin K, Larcher Y, Gelvin MJ, Knue M, Ramos C, Zhao X, Davis SJ, Kim SI, Ranjith-Kumar CT, Choi YJ, Hallan VK, Chattopadhyay S, Sui X, Ziemienowicz A, Matthysse AG, Citovsky V, Hohn B, Gelvin SB (2003b) Identification of *Arabidopsis* rat mutants. Plant Physiol 132:494–505

Ziemienowicz A, Tinland B, Bryant J, Gloeckler V, Hohn B (2000) Plant enzymes but not *Agrobacterium* VirD2 mediate T-DNA ligation in vitro. Mol Cell Biol 20:6317–6322

Ziemienowicz A, Merkle T, Schoumacher F, Hohn B, Rossi L (2001) Import of *Agrobacterium* T-DNA into plant nuclei: two distinct functions of VirD2 and VirE2 proteins. Plant Cell 13:369–384

Ziemienowicz A, Tzfira T, Hohn B (2008) Mechanisms of T-DNA integration. In: Tzfira T, Citovsky V (eds) Agrobacterium. Springer, New York, pp 395–440

Zupan J, Citovsky V, Zambryski PC (1996) *Agrobacterium* VirE2 protein mediates nuclear uptake of ssDNA in plant cells. Proc Natl Acad Sci USA 93:2392–2397

Index